Ion Beam Treatment of Polymers

Ion Beam Treatment of Polymers

Application aspects from medicine to space

Alexey Kondyurin
and
Marcela Bilek

ELSEVIER

Amsterdam • Boston • Heidelberg • London • New York • Oxford
Paris • San Diego • San Francisco • Singapore • Sydney • Tokyo

Elsevier
The Boulevard, Langford Lane, Kidlington, Oxford OX5 1GB, UK
Radarweg 29, PO Box 211, 1000 AE Amsterdam, The Netherlands

First edition 2008

British Library Cataloguing in Publication Data
A catalogue record for this book is available from the British Library

Library of Congress Cataloging-in-Publication Data
A catalog record for this book is available from the Library of Congress

ISBN: 978-0-08-044692-9

Printed and bound in Slovenia

08 09 10 11 10 9 8 7 6 5 4 3 2 1

Contents

Introduction

This book is devoted to modern and future applications of ion beam implantation technologies for polymer materials. Modern ion beam sources have become cheaper and much more effective for use in industrial processes. This has given us the impetus to collect data on modification effects in polymer materials and to show different kinds of polymer processes where the ion beam implanter can offer new properties or solve technological problems. We considered different kinds of applications from kitchen appliances such as dishwashers to high tech devices such as human organism implants or space constructions.

The book is written mostly from our original results (School of Physics at the University of Sydney, Australia; Rossendorf Research Center and Institute of Polymer Research, Dresden, Germany; and Institute of Technical Chemistry, Perm, Russia), and previous experience in ion beam implantation of polymers as well as basic physical and chemical knowledge related to this field are presented. We do not pretend that the scientific description of polymers and industrial polymer processes for ion beam technologies is by any means complete, and if readers can find other useful applications for ion beam implantation of polymers or solutions to any problems using ion beam implantation, we will be happy to hear from you.

Due to our attempts to present ion beam techniques for industry, we do not consider very high energy particle sources because of their cost. Our aim therefore is to concentrate on low cost devices and systems which could be used even in small enterprises for production of special kinds of polymer material and polymer devices, where the ratio of "cost of equipment"/"cost of material" is not too high.

Research into high energy and big accelerators is changing and it is now more common to find small and effective machines being used in research centers and universities. Researchers are now focusing on small ion beam implanters such as plasma immersion ion implanters (PIII) or Kaufman sources, which can be bought using a post-doctoral grant, for example. This, however, limits significantly the ion beam systems by ion energy and ion current. Such systems and their application are the subject of this book.

The book is written using our own experimental and theoretical results as well as known literature data. We appreciate the help of our friends and colleagues, who contributed a great deal of knowledge and effort: Yuri Klyachkin, Valery Begishev, Nail Valeev, Viktor Trushnikov, Gennady Mesyats, Nikolay Gavrilov, Elsa Tereshatova, Sergey Lisenko, Valery Karmanov, Svetlana Astafyeva, Valentina Romanova, Irina Osorgina, Rustam Khaibullin, Reihard Guenzel, Manfred Maitz, Edgar Richter, Emily Pecheva, Galina Nechitailo, Alexander Mashinski, Vladimir Briskman, Stephane Langloise, Manfred Stamm, Pilip Volodin, Jan Weber, Bee Kwan Gan, David McKenzie and Irina Kondyurina.

We also appreciate the help of Helen Smith for language corrections.

1 Interactions of ion beam with polymer: physical picture

Ion beam treatment of a solid target causes a significant transformation to the structure and properties of the treated surface [1, 8]. These changes in the target depend on: its temperature and the material used; the gaseous environment; and the type of implanting ions, their kinetic energy, and the density of ion flux. The first experiments using charged particles to penetrate solids were done by Rutherford. As the ion penetrates the solid target it collides with atoms of the target material. These collisions cause the atoms and electrons to shift slightly from their equilibrium positions and to generate excited phonons which can dissipate energy as they propagate. Atoms and electrons can leave these positions in the target if the transferred energy from the penetrating ion is higher than the ionization energy of the target atom (for the electron) or energy of bonding in the molecule (for the atom). If a recoiled atom or electron has enough kinetic energy, it can transfer energy upon hitting other atoms and electrons. This produces a cascade of collisions. All defects to the structure caused by the displacement and recoiling of atoms and electrons form a field in the target known as the "spur" of the penetrating ion. Usually, the volume of the spur has the shape of a drop: narrow enter, wide waist and obtuse end.

The collision events of flying ions in the target can be considered based on the theory of scattering particles. The energy lost by the penetrating ion at collision with the atom in the target depends on the angle of ion attack, interaction of ion with atom and electron, and density of the target. If we assume that the excitation processes of the electron and atom are not correlated, the energy transfer is the sum of electron and nuclear stopping effects:

$$\frac{dE}{dx} = N[(S_n(E) + S_e(E)] \tag{1.1}$$

where $S_n(E)$ and $S_e(E)$ are nuclear and electron cross-sections of the stopping effects and N is the atomic density of the target [2, 8]. Typically all models used to calculate ion penetration are based on such additive assumptions. The nuclear and electron cross-sections depend on interaction between collided particles. Usually, pair potentials for collided particles are used as WHB or ZBL potentials, which are used in modern models and computer codes for ion collision calculation. Theoretical results of modern computer simulations correspond well with experimental data on ion penetration depth, defect distribution, phonon distribution, scattered atoms and electrons, and transparent ion calculations. Generally, the codes TRIM and SRIM, based on the Monte Carlo method of calculation, are used for the simulation of ion implantation effects in solids, including polymers [2].

For example, an area of nitrogen ion track in polyethylene calculated with the TRIM code is presented in Fig. 1.1. The ion penetrates and collides with carbon and hydrogen atoms, causing them to recoil. With high energy, the recoiled atoms fly apart and collide with other carbon and hydrogen atoms. A tree of collisions grows. Thousands of ions penetrating into randomly distributed target atoms are calculated and analyzed (Fig. 1.2). After simulation a complete statistical analysis of all collisions, stopped ions, recoiled and displaced atoms and electrons as well as phonons is presented. Stopped ions are distributed with maximum in under surface layer (Fig. 1.3). The profile of stopped ions was analyzed by experimental methods for different materials, including polymers, and a good agreement with theoretical

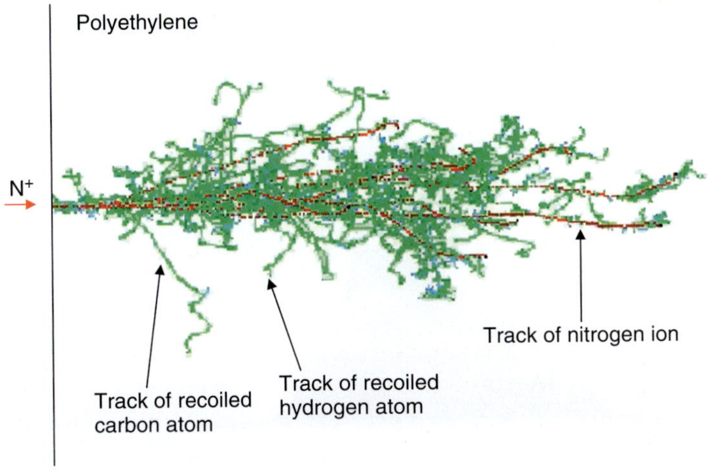

Fig. 1.1. Result of calculation by TRIM codes for ten penetrating ions. Target – polyethylene, penetrating ions – nitrogen, ion energy – 20 keV.

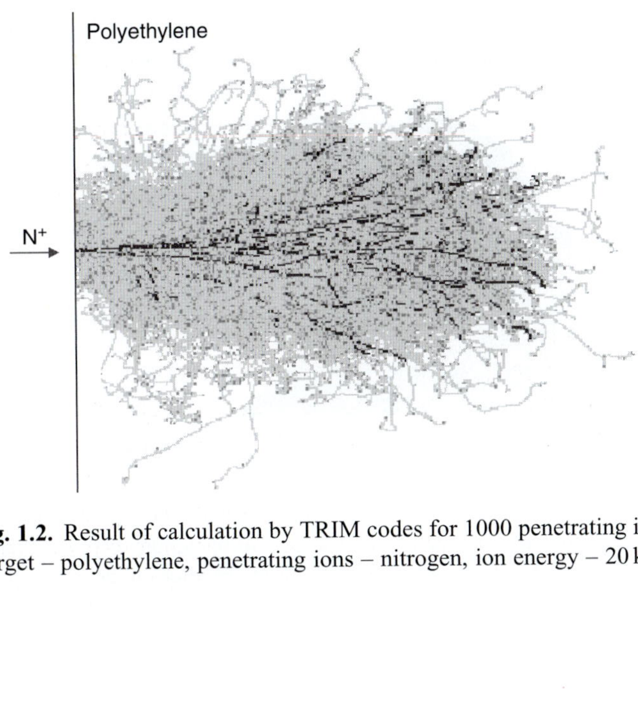

Fig. 1.2. Result of calculation by TRIM codes for 1000 penetrating ions. Target – polyethylene, penetrating ions – nitrogen, ion energy – 20 keV.

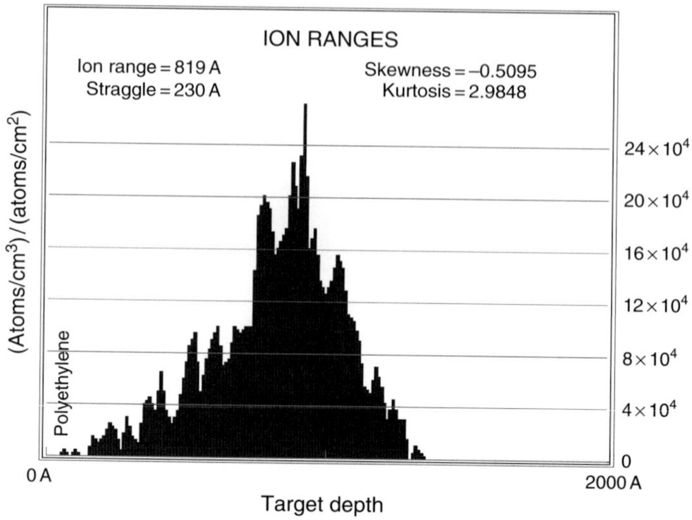

Fig. 1.3. Result of calculation by TRIM codes for 1000 penetrating ions. Target – polyethylene, penetrating ions – nitrogen, ion energy – 20 keV. Nitrogen ion distribution after complete stopping.

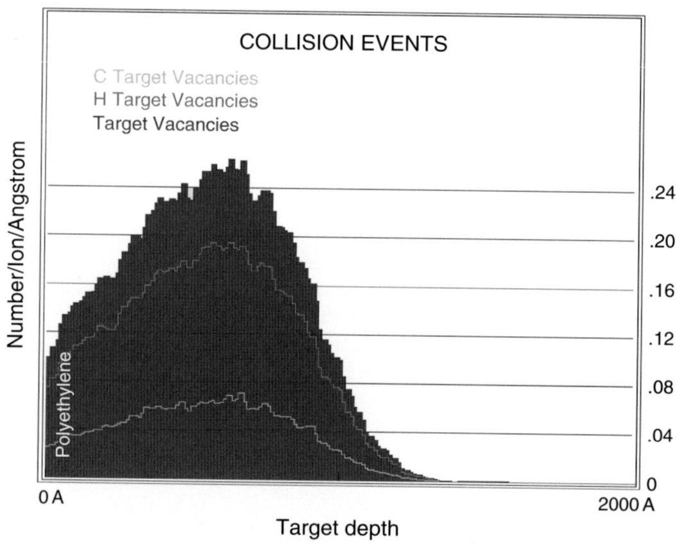

Fig. 1.4. Result of calculation by TRIM codes for 1000 penetrating ions. Target – polyethylene, penetrating ions – nitrogen, ion energy – 20 keV. Distribution of carbon and hydrogen vacancies after ion penetration.

data is observed. For polymer materials a low fluence of implantation is usually used and the distribution of stopped ions is not very important.

More important information is contained in a graph of collision events, which shows the distribution of carbon and hydrogen atom vacancies (Fig. 1.4). This distribution corresponds to a distribution of free valences in polymer macromolecules, or in another words, the distribution of free radicals created by the propagating impacts. The free radicals are a result of significant structural damage to the polymer macromolecule, and initiate complex structural transformations. The sum target vacancies graph cannot be applied to polymers, because the recoiling of carbon and hydrogen atoms has different consequences for structure transformation. Ideally, if a hydrogen atom is recoiled, a carbon atom binds an unpaired electron which is strongly attracted to pair with another unpaired electron. However, the recoiled carbon atom (for example, in a polyethylene macromolecule) generates four unpaired electrons: two at the hydrogen atoms and two at the neighboring carbon atoms. Therefore, to analyze the full number of free radicals, the structure of the polymer macromolecule must be taken into account.

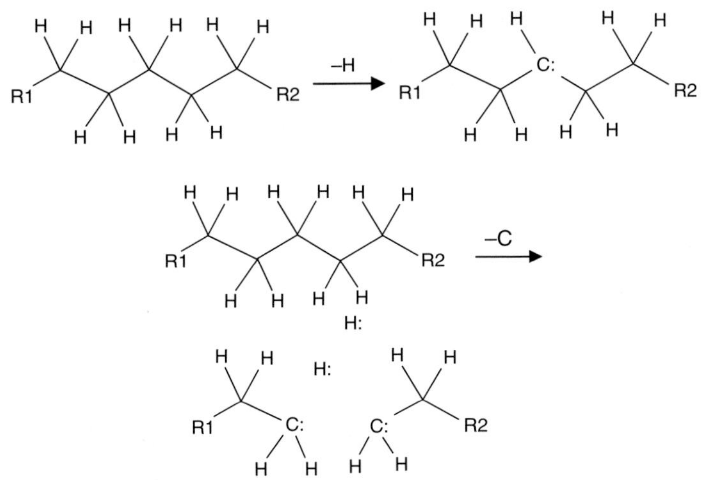

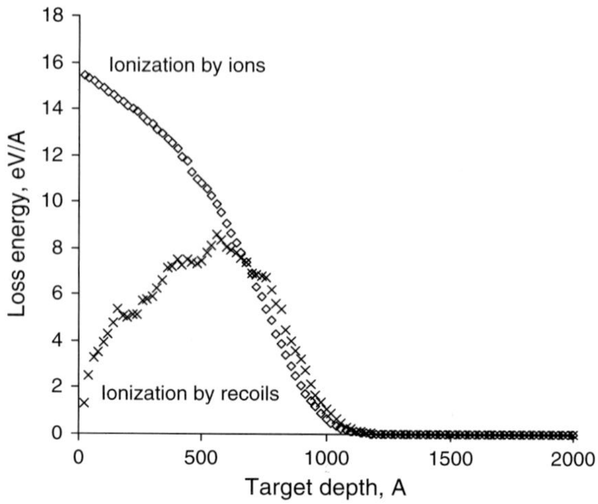

Fig. 1.5. Result of calculation by TRIM codes for 1000 penetrating ions. Target – polyethylene, penetrating ions – nitrogen, ion energy – 20 keV. Energy transfer to ionization of the target atoms caused by penetrating ions and recoiled target atoms.

The density of free radical concentration in polymers is higher than when calculated from ion penetration effects.

The example of energy lost on ionization is presented in Fig. 1.5. The curve of ionization caused by the incoming ion starts from the surface. The penetrating ion causes more ionization effects than produced by the recoiled target atoms. Free electrons appearing due to ionization events can leave the polymer that gives additional charge to the polymer target. Such electrons are called secondary electrons. Usually, the ion generates a large number of spattered electrons and a charge caused by an incoming ion is smaller than the charge caused by electron release. Also free electrons with high energy can penetrate into layers of polymer target that are deeper than the penetration depth of the ion. The collisions of such electrons with polymer macromolecules can cause structure transformation in deep layers.

Phonon distribution is responsible for low energy transfer (Fig. 1.6). A small amount of phonons are generated

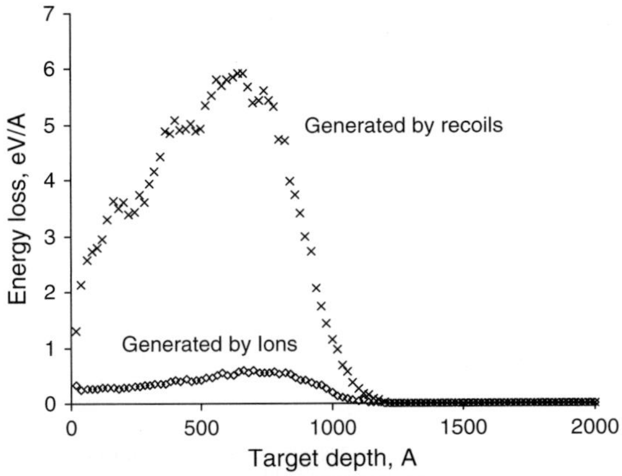

Fig. 1.6. Result of calculation by TRIM codes for 1000 penetrating ions. Target – polyethylene, penetrating ions – nitrogen, ion energy – 20 keV. Phonon distribution generated by ions and recoiled atoms.

by implanting ions and the majority of phonons are generated by recoiled atoms of the target. Phonon excitation can be interpreted in thermodynamical terms as the vibrational temperature of the macromolecules. Recalculation gives the temperature increase up to 10^4 K, which is impossible for polymer materials. After a short time (100 ns) the temperature drops to initial temperature due to phonon dissipation. Local overheating occurs in a very short time. If current density is high enough (it is observed for pulse regimes), the second ion enters the local region of the polymer heated by the first ion (Fig. 1.7). This means that local overheating of the surface layer is expected, especially at the pulse regime of ion implantation with high pulse current density. The overheated region lies under the surface layer and depends on the penetration depth of the ion.

Overheating of the polymer sample can occur at average high current density and weak thermoconductivity of the sample holder. Due to the high sensitivity of polymers, this overheating effect is frequently observed for polymer film in the absence of good contact with the cooled substrate. The film could be wrinkled or even rolled under the ion

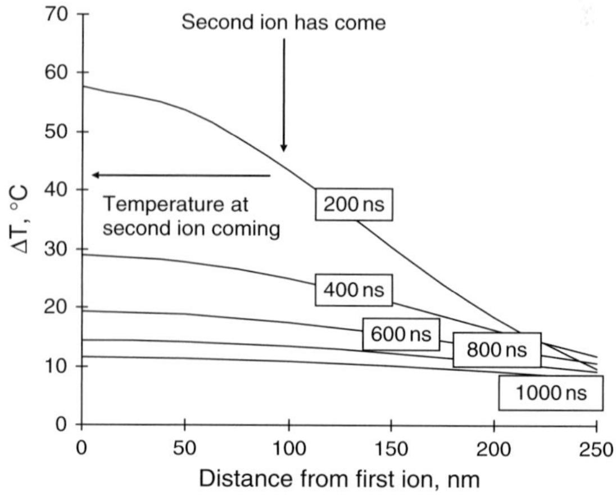

Fig. 1.7. Temperature dynamic after first nitrogen ion penetration with 20 keV into polyethylene. Second ion comes in polyethylene heated by first ion. The calculation is done for a pulse current density of 5 mA/cm^2.

beam. Also the recrystallization of the polymer film can occur if the ion beam density is sufficient. Usually, ion beam implantation is used in regimes of low current density (continuous current density or average if the pulse regime is lower, e.g. 1–10 μA/cm^2), when the overheating regime is not achieved.

After collisions in the thin surface layer of the implanted polymer a wide number of gaseous products are generated. Contamination of the gas atmosphere depends on initial structure of the material. These gaseous products can diffuse into the deep layer of the polymer, or diffuse towards the polymer surface where they are released into the vacuum. For example, the analysis of gaseous products released from polyethylene are typically hydrogen as well as molecular fragments of polyethylene macromolecules: examples include CH_4, C_2H_4, C_3H_6 and others. If the ion current density is high enough, the released gaseous products interact with the ion beam and form a plasma cloud, which is observed by irradiation of light over the treated polymer surface [9]. The spectra of irradiated light can be used to determine the contamination of released products. For example, if nitrogen ion beam is applied to polyethylene, a blue light is viewed, which corresponds to the spectra of nitrogen/hydrogen collisions – the main product released. This is observed at low fluence of ion beam implantation. At high fluence, complete carbonization of the polyethylene surface occurs and carbon is the main product: the irradiating light becomes red which corresponds to the spectra of nitrogen/carbon collisions. The plasma cloud which hangs above the polymer surface (as the result of secondary electron release) can be a conducting media for surface discharge.

The TRIM and SRIM calculations give a very important value: thickness of the modified layer, which is determined as the projective depth of the target vacancies distribution [2]. This layer is mostly modified and associated with the main structure changes of the polymer after ion beam implantation. A correlated characteristic with projective depth of the target vacancies distribution is the projective range, determined as the penetration depth of implanting ions. The projective range may be calculated by TRIM and SRIM codes.

Using the TRIM and SRIM calculation results, we must keep in mind that the calculations are done assuming a low fluence treatment and that the structure of the implanted target does not change during ion penetration. In reality, such an assumption is true at fluences lower than 10^{12}–10^{13} ions/cm^2. At higher fluences the result of ion penetration and radiation effects must be corrected due to the changes in target structure under the ion beam.

References

1. W.A. Grant, J.L. Whitton, R.P. Procter, *Surface Modification of Metals by Ion Beams*, Series: MSE Ser., Elsevier Science, 1987, 438 pp.
2. *Handbook of Ion Implantation Technology*, Edited by J.F. Ziegler, Elsevier, 2003.

3. V.B. Odzhaev, I.P. Kozlov, V.N. Popok, D.B. Sviridov, *Ion Implantation of Polymers.* Minsk: Belorussian State University, 1998.
4. J.J. Cuomo, S.M. Rossnagel, H.R. Kaufman, *Handbook of Ion Beam Processing Technology*, William Andrew Publishing, 1989, 438 pp.
5. A.V. Byeli, V.A. Kukareko, O.V. Lobodaeva, I.I. Taran, S.K. Shikh, *The Ion-Beam Processing of Metals, Alloys and Ceramic Materials*, Physical-Engineering Institute of NASB, Minsk, Belarus, 1998, 220 pp.
6. *Ion Implantation and Beam Processing*, Edited by J.S. Williams and J.M. Poate, Elsevier Science & Technology Books, 1984, 419 pp.
7. *Handbook of Plasma Immersion Ion Implantation and Deposition*, Edited by A. Anders, Wiley-Interscience, 2000, 672 pp.
8. *Fundamentals of Ion-Irradiated Polymers*, Edited by D. Fink, Springer, 2004, 404 pp.
9. M. Ogawa, U. Neuner, T. Endou, S. Yamauchi, M. Nakajima, K. Horioka, Y. Oguri, Plasma-beam interaction experiments at TIT, Poster on BEAMS' 96, Prague, 1996.

2 Sources for ion beam treatment

Ion beam implantation needs an ion beam source (ion beam accelerator). The ion beam source generates a flux of ions with high energy (speed) in a low pressure atmosphere. The ion beam source includes an ion source (plasma discharge, spattered target) which produces a number of ions. These ions are accelerated in an electrical field. The energy of the ions is higher than the kinetic energy of the plasma particles or atoms in the ion source.

Initially, the development of ion beam sources was brought about by experiments with atomic structure (Rutherford's experiments). In the first experiments radioactive isotopes releasing high energy charged particles were used. Later, accelerators based on an electrical field were developed. In 1931 Van-der-Graaf built an electrostatic generator, which is still used for high energy ions. In 1932 a cascade generator was developed which produced a 1 MeV proton beam. Also in 1932 the first nuclear reaction with lithium bombarded by artificially accelerated protons was carried out. Then (1932–1944) accelerators based on cyclotron resonance were developed for 10–20 MeV proton energies. After 1950 accelerators were created based on the phase resonance principle and energies up to some 100 GeV are now available. Such accelerators are used in high energy physics for electron or proton acceleration. The special development of ion accelerators for heavy ions was carried out in the 1950s for the division of uranium isotopes for the nuclear industry. Such accelerators were used for material modification including polymers (Los Alamos Laboratory, Sandia Laboratories, Kurchatov Institute of Nuclear Physics) in the 1960s–1970s. In the 1970s–1980s new ion beam sources were developed for industrial processes of metals and semiconductors. Ion beams with high current density were applied for the modification of alloys, steels, glasses, and for doping technologies of silicon and germanium wafers. Such sources were starting to use polymer modification. However, the cost of ion beam accelerators was too great for industrial application and polymers were treated more for purely scientific investigations. In the 1990s the price of ion beam sources decreased in line with an increase in ion beam diameter, and the use of ion beam implantation is now profitable for polymer materials.

We do not consider all types of accelerators used for the ion beam implantation of polymers – some detailed reviews of ion beam accelerators can be found in the literature. We will, however, concentrate on ion beam sources used for industrial applications related to polymer materials. There are two types of ion beam sources frequently used for polymer modification: ion beam implanters and plasma immersion ion implanters.

The first method, often referred to as conventional beam line ion implantation, relies on the use of biased grids to accelerate a beam of ions out of a plasma source (Fig. 2.1). There are two kinds of ion beam implanter: continuous and pulse ion beam. In first case the accelerating field and plasma source works continuously, due to space charge

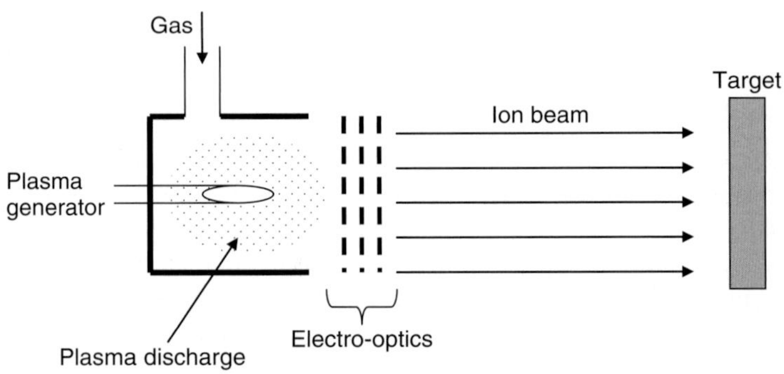

Fig. 2.1. Simple scheme of ion beam implanter.

effects in the beam; the ion currents are typically limited to the range of microamperes. In second case, the pulsed accelerating electrical field with continuous plasma source or continuous accelerating electrical field with pulsed plasma source generates a high current ion beam in a pulse-periodical regime. The current density during a short pulse could be some A/cm^2 but the average current density is some μA/cm^2.

In all these cases, ion implantation is achieved by placing the material to be modified in the path of the extracted ion beam. The method is intrinsically line of sight so that uniform treatment of complex 3D surfaces can only be achieved by mechanically rotating the object so that the ion beam "paints" over the entire surface. The advantages of this method include the straightforward selection of a monoenergetic ion beam of a single species for implantation; ion energy could be up to MeV. Disadvantages of this method include the complexity of the motion required to give a uniform treatment over the surface of complex 3D forms.

An example of the first kind of accelerator is the "Pulsar" ion beam accelerator of the Institute of Electrophysics, Ekaterinburg, Russia. The accelerator includes a plasma source based on arc discharge or glow discharge plasma with a hollow cathode in a magnetic field. An electro-optic system with three high voltage electrodes produces ion beams from plasma discharge volume. The accelerator generates a beam of gaseous ions (N_2, O_2, Ar, C_3H_8 and others) with energy from 2 keV to 40 keV. Cross-section of the beam is 100–200 cm^2 at current density deviation on a beam diameter no higher than 10%. The regime of beam generation is pulsed, pulse duration is 0.03–1 mS, and pulse repetition frequency is 0.1–100 Hz. The pulse current density could vary between 10 μA/cm^2 and 10 mA/cm^2. The source is jointed with a vacuum chamber which could operate at low pressure (up to 10^{-3} Pa). The source is compact and used in technological processes.

The second method of supplying energetic ions to a surface is known as plasma immersion ion implantation (PIII)

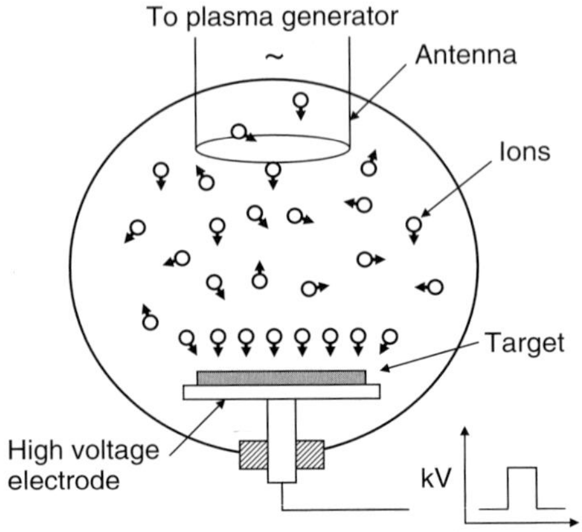

Fig. 2.2. Plasma immersion ion implanter.

and has many advantages as a method for ion modification of polymers. It is more cost effective than beam line implantation, providing higher fluxes of ions per unit time and eliminates the need to rotate the object being implanted to achieve modification of a three-dimensional surface. But the energy of ions is limited to 50–60 keV. The PIII method was developed in the late 1980s [5] to implant nitrogen into steel surfaces for hardening. The basic principle is to immerse the object to be treated (workpiece) directly into a plasma and to bias it with high voltage (in the kilovolt range) to draw ions out of the plasma and accelerate them into the surface of the workpiece (Fig. 2.2).

An example of the second kind of ion beam source is the plasma immersion ion implanter of the School of Physics, University of Sydney, Australia (Fig. 2.3). This setup is a home-made machine and is basically a three part vacuum chamber: stainless steel chamber, glass chamber and aluminum chamber. The inductive-coupled antenna for plasma power is maintained on the glass part of the chamber. A vacuum up to 10^{-4} Pa is provided by a turbo-molecular pump and rotary pump. A gas flow controller provides stable pressure of the working gas in the chamber. The radiofrequency plasma generator (13.75 MHz) provides plasma power of

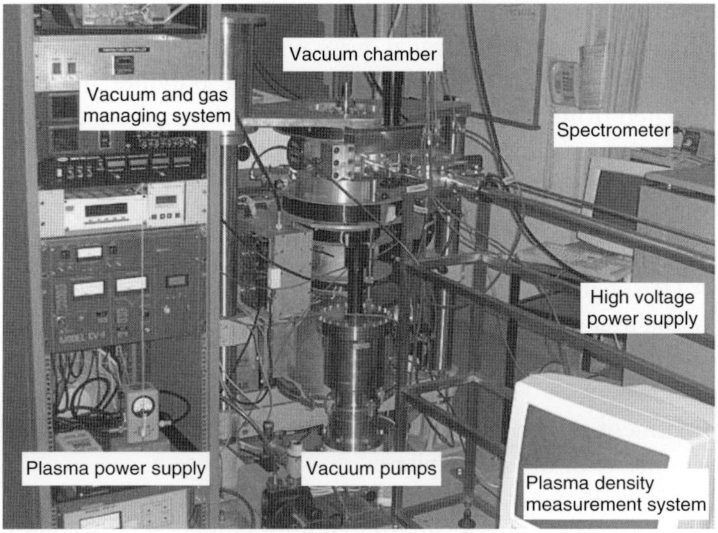

Fig. 2.3. Plasma immersion ion implanter of the School of Physics, University of Sydney.

50–400 W through an antenna in the chamber. Stable plasma power is provided by matching the wave resistance of the plasma generator with the wave resistance of the plasma discharge using a “matching box”. The aluminum part of the chamber is attached by magnetic coils which create a magnetic field in the chamber. Plasma density can be maintained from 10^{14} to 10^{16} ions/cm^3. A high voltage electrode is inserted into the chamber. High voltages from 1 keV to 40 keV are applied in a pulse regime: pulse duration from 5 to 100 μs at a pulse frequency of 20–400 Hz. High voltage pulses are provided by a power supply produced by the Australian Nuclear Center ANSTO. Operation of high voltage power supply could be managed by computer. The implanter parameters are monitored by an Ocean Optic spectrometer for analysis of irradiation spectra of plasma discharge and by a Langmuir probe for plasma density analysis.

A polymer sample is placed on a high voltage electrode in a vacuum chamber. The area of the polymer sample could be up to 150 mm in diameter.

The polymer sample could be manipulated in the vacuum chamber by a mechanical arm without contact with the atmosphere. This is important when the polymer must be isolated from atmospheric contact. With the mechanical arm the polymer sample can be changed through an air load lock without opening the vacuum chamber to the atmosphere. The polymer sample can be moved and placed on ATR crystal of an FTIR spectrometer under vacuum and FTIR ATR spectra of the sample can be recorded without contact of the treated sample with the atmosphere. Also, the polymer sample can be post-treated by aggressive media in chemical attachment isolated from the vacuum chamber. The post-treatment can be done without contact of the treated polymer sample with the atmosphere.

The PIII system can be used in a convenient plasma chamber. In some cases the additional high voltage power supply and high voltage electrode installed in plasma chamber could be enough for transformation of the radio-frequency plasma chamber into the PIII source. This affords the opportunity of getting a PIII source of high voltage ions without the great expense associated with using an ion beam implanter.

The ion beam implantation process of polymer differs in the case of ion beam source and plasma immersion ion implantation. In the case of ion beam source, the beam is generated separately from the polymer target. The presence of the non-conducting polymer target does not influence on the ion beam. In the case of plasma immersion ion implantation, the ion flux is generated in space near the non-conducting polymer target. The presence of the polymer target influences the electrical field near the polymer surface and changes the ion beam formation.

As soon as the bias is applied to the high voltage electrode, a sheath forms around the workpiece as electrons are repelled from its boundaries on a nanosecond timescale. This leaves a space charge of ions, which shields the rest of the plasma from the high bias, close to the workpiece. The ions' higher inertia prevents them from moving on timescales typical of electron motion. The sheath which exists on these short timescales is known as the *matrix sheath* [1] (p. 115) because the ions maintain the "matrix" associated with their locations in the quasi-neutral plasma prior to electron motion. The matrix sheath forms around the workpiece and is more or less conformal to its shape. The width of this sheath is given by:

$$S_M = \left(\frac{2\varepsilon_0 V_0}{en}\right)^{1/2} \tag{2.1}$$

where ε_0 is the permittivity of free space, e is the electronic charge, V_0 is the bias potential applied to the workpiece and n is the plasma density. A strong electric field drops across the sheath. Ions, already located within the sheath as well as those that drift into the region from outside, are accelerated towards the workpiece. The timescale for ion motion (the inverse ion plasma frequency), given by:

$$\tau_i = 2\pi \left(\frac{\varepsilon_0 M}{e^2 n}\right)^{1/2} \tag{2.2}$$

where M is the ion mass, is typically in the range of microseconds.

In the absence of collisions, the ions gain kinetic energy equal to the potential through which they fall on their way to the workpiece. They then implant into its surface with energies up to their charge multiplied by the applied bias voltage. As the ions are accelerated and implanted into the workpiece their density in the sheath is reduced and the sheath must expand in order to contain enough positive charge to continue to screen the applied bias from the bulk plasma. This process continues until an equilibrium density profile, consistent with the acceleration of ions in the sheath, is formed. Equilibrium occurs when the ion current entering the sheath is equal to the space charge limited current that flows through it. At this point the sheath is known as an *equilibrium* or *Child-law* sheath [1] (pp. 115–117) and its width on a planar workpiece is given by:

$$S_{CL} = \frac{2}{3}\sqrt{\varepsilon_0}\left(\frac{2e}{M}\right)^{1/4}\frac{V_0^{3/4}}{\sqrt{env}} \quad (2.3)$$

where v is the component, normal to the sheath/plasma boundary, of the velocity with which the ions enter the sheath. In a non-drifting plasma, v is the Bohm speed given by $\sqrt{eT_e/M}$, where T_e is the plasma electron temperature in electron volts. For plasmas, such as created in cathodic arc or laser ablation, v is the greater of the ion drift component normal to the sheath/plasma boundary and the Bohm speed.

Typical laboratory plasmas range in density between 10^{13} and $10^{17}\,\mathrm{m}^{-3}$, with matrix sheath dimensions for a planar workpiece biased at 10 kV, ranging from 30 cm to 3 mm, respectively, and Child-law sheath dimensions ranging from 1.4 m to 1.4 cm, respectively. Clearly for the low plasma density range where a background gas is present, collisions will not be negligible. Any collisions with neutrals in the sheath will reduce the energy with which the ions impact the workpiece. Such effects must be taken into account when determining the depth of ion modification achieved.

Both drift velocities [4, 2] of ions greater than the Bohm speed and substrate curvature [3] act to reduce the Child-law sheath dimensions. Both of these effects are important to note because they place limits on the maximum plasma density, in which PIII with a particular bias can be applied, for a given workpiece geometry and plasma drift geometry. If the sheath becomes too thin in any one location, the electric field strength will exceed the vacuum breakdown limit and arcing across the sheath will result [2]. Aside from interrupting the ion implantation process, arcs can also cause damage through ablation of the workpiece surface and must be avoided.

In the absence of collisions in the Child-law sheath, the energies of implanted ions are typically higher than those implanted from the matrix sheath since all implanting ions are accelerated through the entire sheath width in the equilibrium phase. In this phase, it is in principle possible to keep implanting ions at a steady rate, their supply limited only by their generation rate in the bulk plasma. Practical limits are, however, imposed by heating and charging of the implanted surface. Both of these effects are much more limiting in the case of polymers than for other workpiece materials.

It is usually necessary to apply the bias in a pulsed mode in order to control the heat load imposed on the workpiece by the implanting ions and to ensure a good supply of ions. Depending on the construction of the plasma source, the ions are often extracted from the plasma at a faster rate than they are generated so the pulse off time is required to replenish them.

The two most important aspects of the ion implantation process for determining the modification achieved in the workpiece surface are the fluence, or total number of ions per unit area implanted and the energy distribution of those ions. The fluence varies linearly with the time over which the PIII process is applied. The spread of energies of the ions impacting on the surface is determined by (1) the shape of the voltage pulses applied, (2) limitations on the voltage appearing on the surface due to the dielectric constant of the workpiece and surface charging and (3) the extent to which ions collide with other species, including background gas molecules, in the sheath.

Since the ion density in the sheath region is highest in the matrix sheath and most of the matrix sheath ions are not accelerated over the whole sheath width, there is a substantial ion flux in any PIII process which has energy below that corresponding to the applied bias.

The proportion of low energy ions implanted for a given bias pulse length will be substantially higher if the pulse rise time is slower than the characteristic response time of the ions as given by equation (2.2). For most currently available pulsed power supplies the rise time is between several nanoseconds and several microseconds so in many practical cases the rise time will exceed the time taken to form the Child-law sheath. In such cases, the spread of ion energies of ions implanted into a conducting surface using the PIII method is primarily determined by the rise time of the applied voltage pulse [7, 13].

For insulating surfaces, such as polymers, the build-up of charge on the surface can be a major problem which if left unchecked reduces the energy with which subsequent ions enter the surface [12]. This can be an issue with both beam line ion implantation and the PIII method. Due to the much higher ion currents implanted using the PIII, it occurs much more rapidly during this process. The surface charge reduces the bias voltage appearing at the workpiece surface and the energies of implanted ions correspondingly. Modified Child-law sheath equations which allow the effect to be estimated are presented by Emmert *et al.* [6].

In the case of polymeric materials, their high secondary electron emission coefficient increases the rate of charging. Because of the electric field present at the polymer surface, secondary electrons are repelled from the surface leaving behind an equivalent net positive charge. As the charge builds up, the potential at the negatively biased surface rises. If the process continues long enough the surface will attain the floating potential [1] (p. 112) (i.e. the potential that an unbiased object immersed in a plasma achieves – usually a few tens of volts).

As the surface potential rises, the potential difference between the workpiece and the plasma is reduced causing a corresponding reduction in the energies of ions implanted into the surface. The reduction in the voltage across the sheath, associated with the build-up of surface charge, also causes a sheath contraction (as per equation (2.3)). This collapse of the sheath has been measured using Langmuir probes by Oates *et al.* [9] for PIII from a filtered cathodic vacuum arc plasma. The measurements showed that applied voltages of several kV were fully compensated on microsecond timescales by the accumulation of surface charge indicating that surface charging is a severe problem for the implantation of polymeric workpieces.

In cases where the workpiece is a polymeric film or membrane with a low dielectric breakdown strength compared to the PIII bias being applied this issue can be neglected. Breakdown from top of polymer to high voltage electrode occurs by two mechanisms: through the film and across the polymer surface. In these cases, continuous cycles of charging and breakdown occur during the bias pulse. This will cause the potential to oscillate between the applied bias and the applied bias less the breakdown strength, with corresponding fluctuations imposed on the energies of implanting ions. The reductions in ion energy associated with voltage droop-induced sheath collapse can be mitigated by use of pulse durations which are very much shorter than the characteristic time of the droop. The flow of plasma electrons to the surface in between pulses serves to neutralize the accumulated charge between pulses. In some cases, the changing effect could not be detected during pulses. For example, the polyethylene film of 20 μm thickness has maximal breakdown strength of 1000 V and maximal variations of bias could be expected less than 5% if applied bias is 20 kV.

The charge on the top of the polymer gives a problem with arcing to the plasma sheath. When the charge on the top of the polymer sample becomes high enough, the electrical field is concentrated between the top of the polymer and the electrode. The electrical field between the top of the polymer and the plasma sheath decreases or becomes zero. When breakdown through the polymer or across the polymer surface occurs, it generates a number of charge carriers (electrons and ions) with high local concentration. After breakdown the electrical field in the polymer becomes zero and the electrical filed between the top of the polymer and the plasma sheath becomes high. However, the presence of charge carriers causes a breakdown between the top of the polymer and the plasma sheath. The breakdown has the characteristics of an avalanche. This is observed as arcing with light emission and high current, and the damping of high voltage. The polymer burns locally and ion fluence

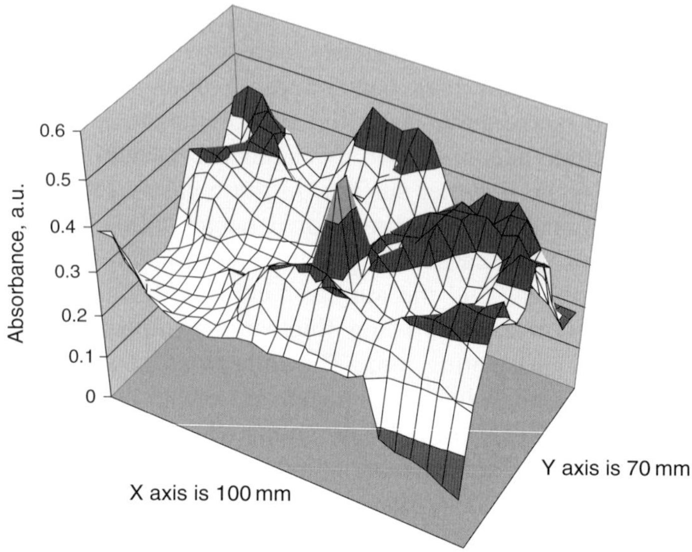

Fig. 2.4. Fluence distribution on polyethylene target after PIII without mesh.

distribution is non-uniform on the polymer surface. Such a strong non-uniform ion beam fluence on the polymer surface is observed in experiments of the PIII treatment of polyethylene film by the mapping of UV spectra (Fig. 2.4). This distribution has stochastic character and depends on individual conductive properties of the polymer sample.

Mesh assisted PIII is a method of considerably reducing the charging problem in polymers [8, 11]. In this approach, a conducting mesh, which is biased in the same way as the workpiece, is placed a small distance in front of the workpiece (Fig. 2.5). The plasma sheath then forms around the mesh and space between the mesh and the workpiece is free of electric fields. The electric fields which accelerate the implanting ions form in the sheath between the mesh surface and the bulk plasma. The ions are then accelerated to the mesh surface and those that pass through the mesh holes continue on to the workpiece. Although the implanted ions still create an accumulating surface charge, the increases in charging rate due to the emission of secondary electrons

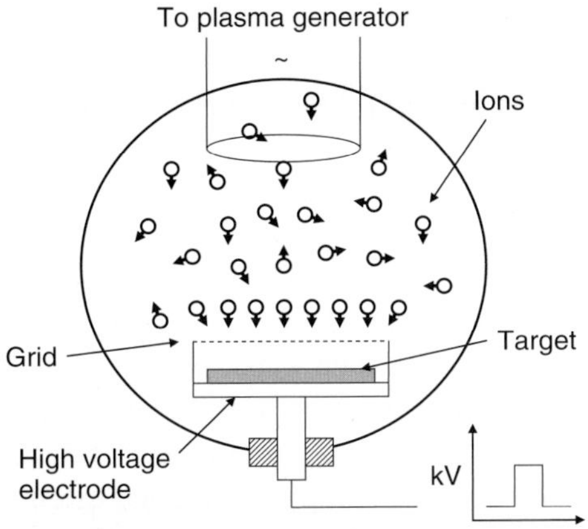

Fig. 2.5a. Plasma immersion ion implanter with additional grid electrode.

Fig. 2.5b. Mesh electrode for PIII. Diameter of mesh is 100 mm, cell of mesh is 0.7 mm and distance from polymer surface is 30–45 mm. Diameter of uniform dose distribution circle on polymer surface is about 50 mm.

are usually avoided. As the surface becomes slightly more positive than the negatively biased mesh, the secondary electrons generated at the polymer surface are attracted back to the polymer resulting in no net charging from this source. This charge compensation from secondary electrons is best achieved with conformal meshes held close to the surface and completely enclosing the workpiece. Disadvantages associated with the use of meshes include surface contamination arising from material sputtered from the mesh. In some cases implanted fluence surface distributions modulate by an image of the mesh, which can be excluded by selection of mesh size and mesh distance from the polymer.

The use of a mesh mitigates the effect of reducing the fluence distribution deviation on the polymer surface. Figure 2.6 shows a map of the UV spectra absorbance for polyethylene after PIII. The mesh provides an area of uniform fluence distribution of polyethylene surface with about 50 mm diameter. This distribution does not depend on individual properties of the sample. The edge effects of fluence distribution are caused by shadowing and focusing of the ion beam by the walls of the mesh cap.

The fluence of implanting ions can be calculated from pulse parameters:

$$F = \frac{j \cdot \tau \cdot f}{e \cdot n} \tag{2.4}$$

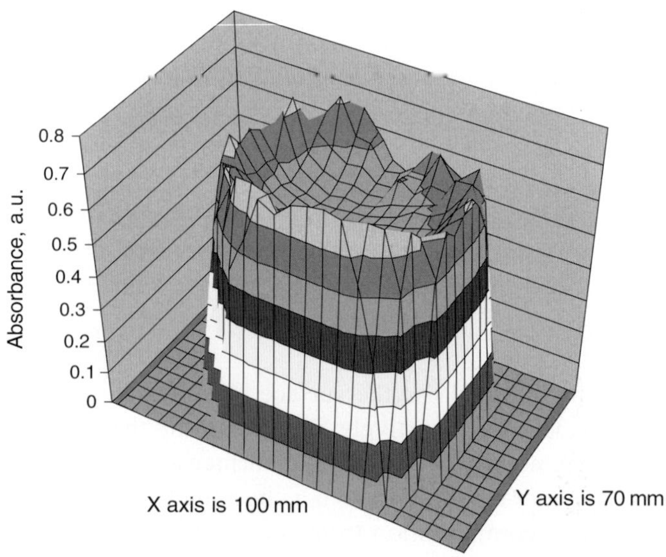

Fig. 2.6. Fluence distribution on polyethylene target after PIII with mesh electrode.

where j is the current density during pulse, τ is the pulse duration, f is the pulse frequency, e is the electron charge and n is the charge number of ion. For an ion beam implanter this formula could be used if all parameters are known. In most cases, the charge number of ion and current density during pulse cannot be calculated from ion source parameters. The current density could be measured by Faraday cap and the ion beam contamination could be determined by mass-spectrometer measurement or by optical measurement of the irradiation light of plasma.

Such monitoring of plasma and ion beam is very useful, because the implantation effect depends strongly on contamination of the chamber environment. Residual gases, especially vacuum oil and greases, can significantly influence the polymer surface layer after ion beam implantation. In spite of this, the change of working gas requires the chamber to be cleaned with plasma, because the previous gas could be implanted into walls of the chamber and 5–10% of residual gas could be observed at the first time of operation after gas change.

A Faraday cap is used for the measurement of current density in ion implanters (Fig. 2.7). The Faraday cap consists of two metal cylinders: the internal cylinder is connected to a measuring device (current integrator or oscilloscope); the external cylinder is under DC potential to exclude the effect of secondary electrons sputtering from the internal cylinder. The ions falling down on the internal cylinder are detected by the measuring device. The current measured in

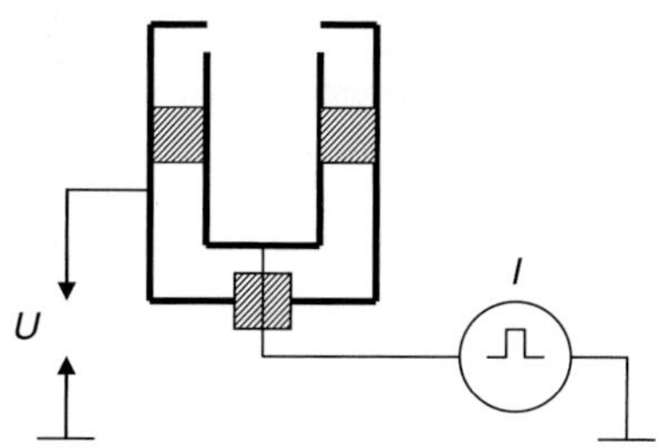

Fig. 2.7. Faraday cap scheme for current density measurements.

the internal cylinder divided by the cylinder bottom area gives current density.

In the case of PIII treatment, the method of indirect measurement of fluence dependence by structure transformation of known polymers is considered in Chapter 4.

In some cases of technological application gas contamination is not important and residual atmosphere (nitrogen and oxygen mixture) is enough for ion beam implantation. For such applications, a rotary pump system is sufficient for evacuating the chamber. For example, an implantation machine scheme with step-by-step vacuuming is presented in Fig. 2.8 for continuous polyethylene film modification. The polyethylene film leaves an outcoming roller and enters a vacuum chamber with rollers providing isolation from atmospheric pressure. Between the rollers the rotary pumps provide a consistent step-by-step increase in vacuum up to the working pressure of ion beam implantation. A number of ion beam implanters are used for the modification of polyethylene. The speed of the polyethylene film running and ion current density from the implanters provide fluence collection on the polyethylene surface. The treated film runs between the rollers with a vacuum pump chain to provide a pressure difference in the outcoming part of the machine.

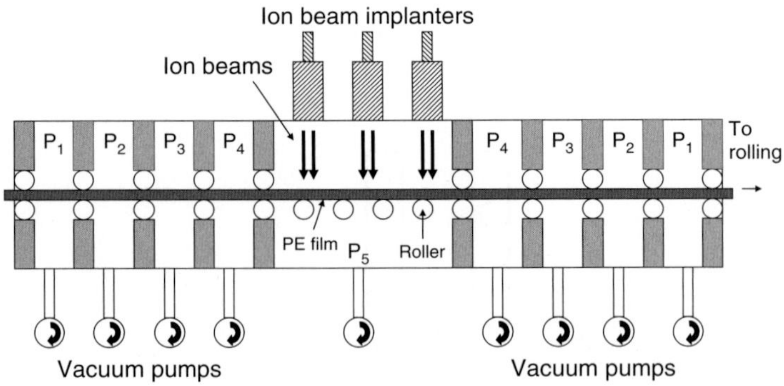

Fig. 2.8. Scheme of ion beam implantation machine for continuous modification of polyethylene film in rolling process.

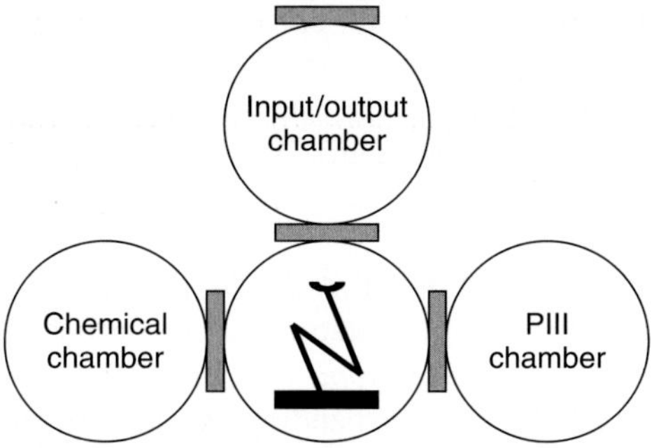

Fig. 2.9. Scheme of PIII chamber assembled with chemical chamber through manipulator unit. Periodical treatment processing of polymer with post-treatment by aggressive chemicals after PIII is possible.

In some cases post-treatment of the polymer modified surface is preferable without contact with the atmosphere. In such cases, a manipulator unit can be used for moving treated polymer into a separate chemical chamber jointed to the plasma chamber. An example of the manipulator scheme is presented in Fig. 2.9. Periodical processing could be applied for polymer surface modification with a robotic manipulator unit. Separating the chemical chamber from the plasma chamber prevents the corrosion effect of the plasma chamber and polymer contamination before modification. The same scheme can be used for processing polymer modification, when the ion beam is also used for sterilization of the polymer surface and the modified polymer is packed in sterilized containers.

References

1. *Handbook of Plasma Immersion Ion Implantation and Deposition*, Wiley, New York, 2000.
2. A. Anders, Width, structure and stability of sheaths in metal plasma immersion ion implantation and deposition: measurements and analytical considerations, *Surf. Coat. Technol.* 136 (2001) 85–92.

3. M.M.M. Bilek, Effect of sheath evolution on metal ion implantation in a vacuum arc plasma source, *J. Appl. Phys.* 89 (2001) 923–927.
4. I.G. Brown, O.R. Monteiro, M.M.M. Bilek, High voltage sheath behavior in a drifting plasma, *Appl. Phys. Lett.* 74 (1999) 2426–2428.
5. J.R. Conrad, J.L. Radtke, R.A. Dodd, F.J. Worzala, N.C. Tran, Plasma source ion-implantation technique for surface modification of materials, *J. Appl. Phys.* 62 (1987) 4591–4596.
6. G. Emmert, Model for expanding sheaths and surface charging at dielectric surfaces during plasma source ion implantation, *J. Vac. Sci. Technol., B: Microelectronics Processing and Phenomena* 12 (1994) 880–883.
7. D.T.K. Kwok, M.M.M. Bilek, D.R. McKenzie, P.K. Chu, The importance of bias pulse rise time for determining shallow implanted dose in plasma immersion ion implantation, *Appl. Phys. Lett.* 82 (2003) 1827–1829.
8. J.N. Matossian, R.W. Schumacher, D.M. Pepper, Surface potential control in plasma processing of materials. Hughes Aircraft Company. 995864[5374456]. 20-12-1994. Los Angeles, CA. 23-12-1992. Ref. Type: Patent.
9. T. Oates, J. Pigott, D. McKenzie, M. Bilek, Electric probe measurements of high-voltage sheath collapse in cathodic arc plasmas due to surface charging of insulators, *IEEE Trans. Plasma Sci.* 31 (2003) 438–443.
10. X.B. Tian, P.K. Chu, Multiple ion-focusing effects in plasma immersion ion implantation, *Appl. Phys. Lett.* 81 (2002) 3744–3746.
11. X.B. Tian, K.Y. Fu, P.K. Chu, S.Q. Yang, Plasma immersion ion implantation of insulating materials, *Surface Coat. Technol.* 196 (2005) 162–166.
12. X.B. Tian, R.K.Y. Fu, J.Y. Chen, P.K. Chu, I.G. Brown, Charging of dielectric substrate materials during plasma immersion ion implantation, *Nucl. Instrum. Methods Phys. Res., Sect. B* 187 (2002) 485–491.
13. X.B. Tian, D.T.K. Kwok, P.K. Chu, Modeling of incident particle energy distribution in plasma immersion ion implantation, *J. Appl. Phys.* 88 (2000) 4961–4966.

3 Interaction of ion beam with polymer: chemical picture

Polymer is a material which is formed by long macromolecules with high molecular mass or network of macromolecules. Polymer macromolecule is formed by jointed monomers – regular unit of the polymer chain. Commonly, the properties of the polymer depend on the type of monomer. An example of the simplest macromolecule is the saturated hydrocarbon macromolecule. A well-known representative of such types of polymer is polyethylene. The polyethylene (PE) macromolecule contains the simplest monomer unit $—CH_2—CH_2—$ which repeats thousands or millions of times in the polymer chain. There are more complex saturated hydrocarbons such as:

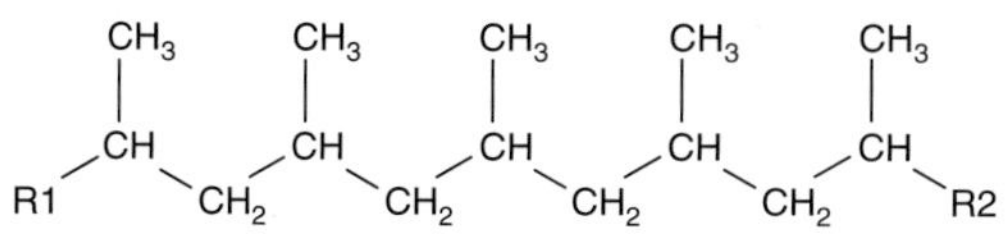

Polypropylene (PP),

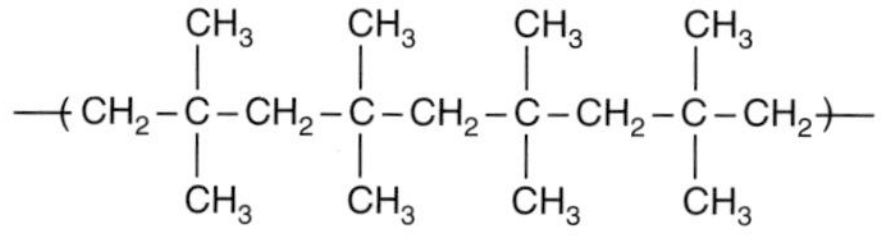

Polyisobutylene

which are homopolymers, because they contain one kind of monomer in the macromolecule. There are heteropolymers, which consist of two and more different monomer units. For example,

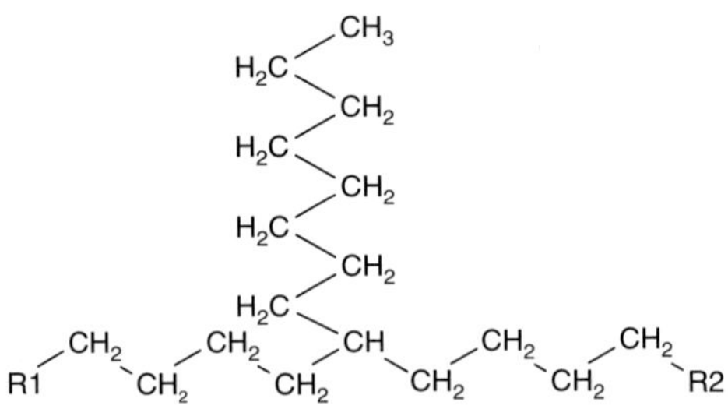

Co-polymer ethylene-octene-1

which contains two kinds of monomers: an ethylene unit and an octane-1 unit.

If a macromolecule contains an unsaturated double or triple carbon—carbon bond, such a polymer is called an unsaturated hydrocarbon polymer, for example:

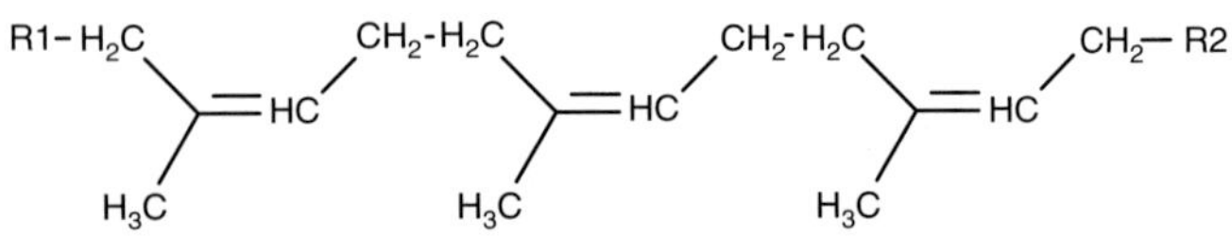

Polyisoprene

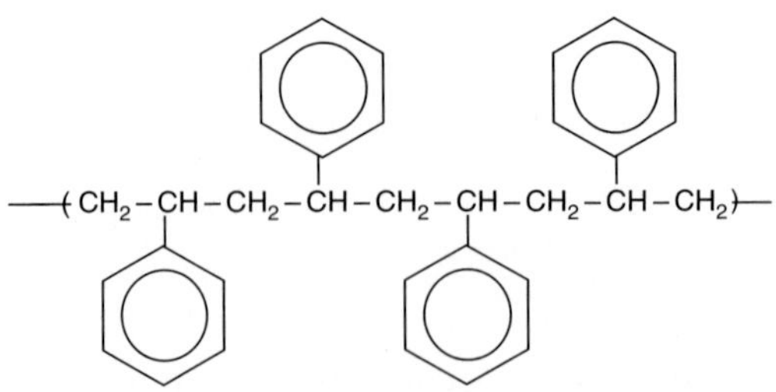

Polystyrene (PS)

Polymer macromolecules can contain heteroatoms (oxygen, nitrogen, phosphorus and others) in the backbone of a side chain:

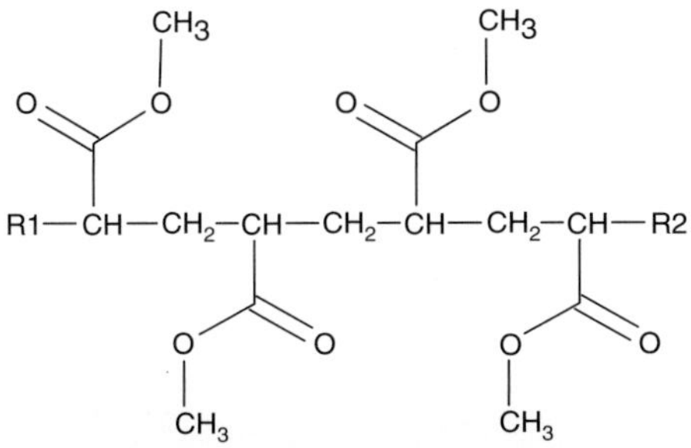

Polymethylemethacrylate (PMMA)

or in a basic (backbone) chain:

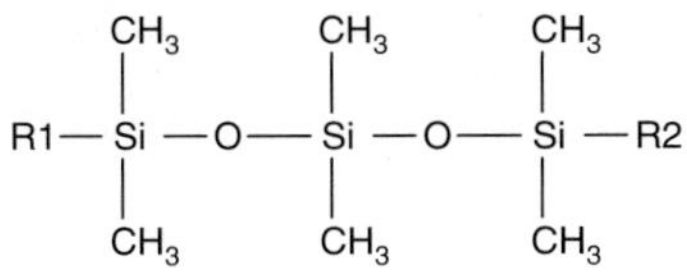

Polydimethylsiloxane (PDMS)

Special kinds of polymers are related to halogen-containing polymers such as:

$$-\!\!\left(CF_2-CF_2-CF_2-CF_2\right)\!\!-$$

Polytetrafluorethylene (PTFE)

Table 3.1 contains a number of polymers with different monomers which were investigated after ion beam implantation.

Macromolecules contain hundreds and thousands of monomers. The mass (or number of monomers, or length) of a macromolecule is characterized by molecular mass distribution function and average molecular mass (MM). Usually the MM of a polymer is in the range 10^4–10^6. Polymers with MM in the range 300–3000 are viscous liquids and are called oligomers.

Table 3.1. Kinds of polymers and their references on ion beam implantation.

Polymer title	Monomer unit	References
Polyethylene (PE), in particular:	$-(CH_2-CH_2)-$	18, 28, 32, 40, 45, 53, 54, 60, 73, 74, 88, 107
• Low density polyethylene (LDPE)		29, 46, 90, 97, 99, 100
• Ultra high molecular weight polyethylene (UHMWPE)		11, 14, 46, 65, 76, 82, 87, 89, 110, 112
Polystyrene (PS) and co-polymers	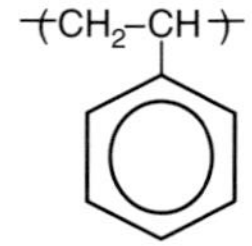	9, 18, 19, 20, 23, 26, 28, 33, 35, 37, 38, 39, 42, 47, 49, 54, 58, 62, 63, 70, 71, 72, 77, 95, 96, 111
Polyethylenterephthalate (PET, Mylar)	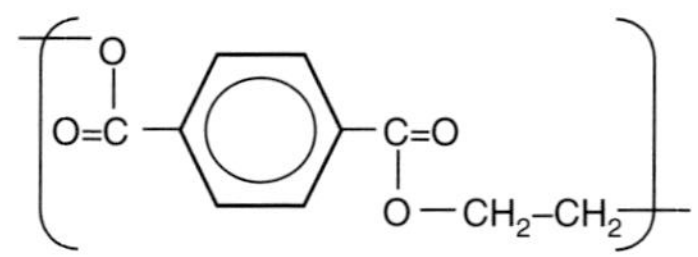	6, 15, 30, 44, 53, 55, 56, 57, 66, 69, 83, 84, 86, 102, 107, 118
Polymethylmethacrylate (PMMA) and co-polymers	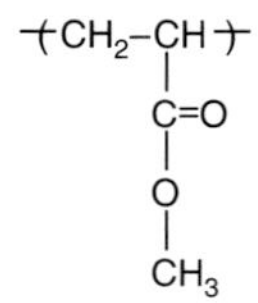	2, 9, 16, 18, 20, 21, 28, 31, 41, 48, 52, 63, 111, 112

Polytetrafluorethylene (PTFE, Teflon) and related co-polymers	$-(CF_2-CF_2)-$	3, 4, 13, 24, 28, 51, 63, 64, 80, 85, 98, 109, 113, 116
Polyimide (PI, Kapton)	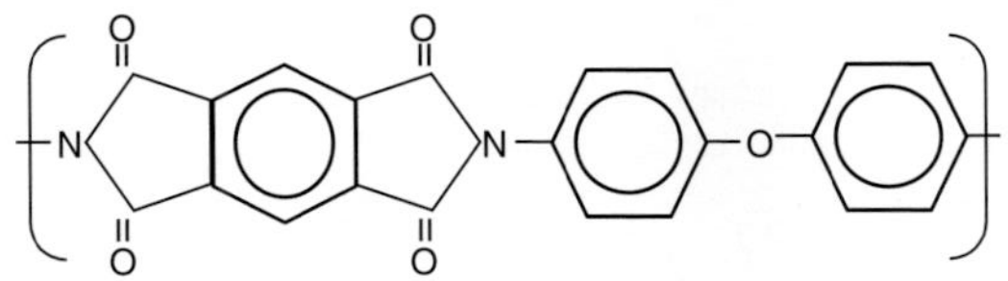	12, 17, 18, 27, 29, 31, 36, 44, 53, 78, 79, 102, 103, 104, 108, 111
Polyurethane (PU)	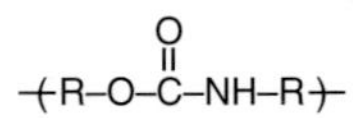	8, 33, 34, 38, 59, 68, 75, 91, 92
Polyethersulfone (PES)	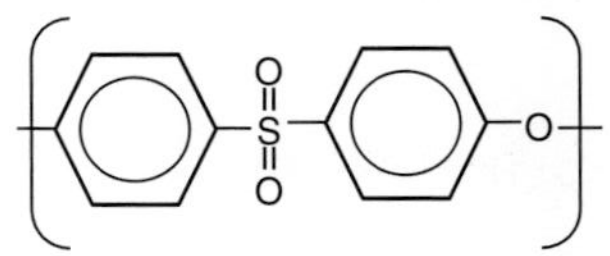	12, 22, 44, 47, 79, 107, 108
Silicone rubber, polysiloxane	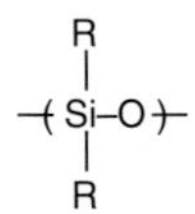	25, 43, 50, 56, 57, 59, 93, 115

(Continued)

Table 3.1. (Continued)

Polymer title	Monomer unit	References
Polyvinyl chloride (PVC)	$-(CH_2-CHCl)-$	7, 9, 47
Polyvinyl fluoride (PVDF)	$-(CH_2-CHF)-$	107
Polyethylene glycol (PEG)	$HO-(CH_2-CH_2-O)-H$	1, 5
Polyamide (PA)	$-(R-C(=O)-NH-R)-$	40, 61, 94, 107
Polycarbonate (PC)	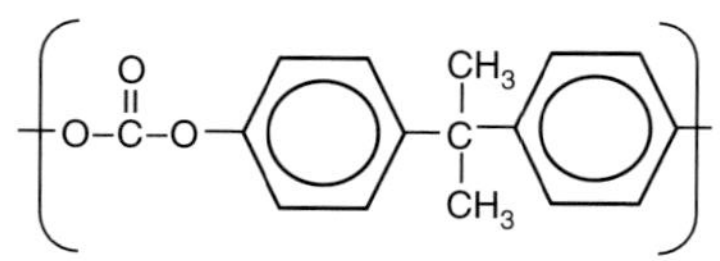	10, 44, 63, 105, 114
Polypropylene (PP)	$-(CH_2-CH(CH_3))-$	53, 106, 107, 112
Polycaprolactone (PCL)	$-(O-(CH_2)_5-C(=O))-$	1, 15

Polyvinylpyridine (PVP)	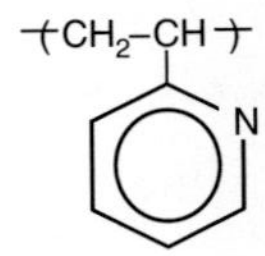	22
Polyvinyl acetate and co-polymers (PVA)	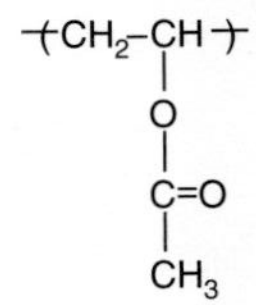	9, 59, 107
Poly glycidil methacrylate-co-3-chlorostyrene	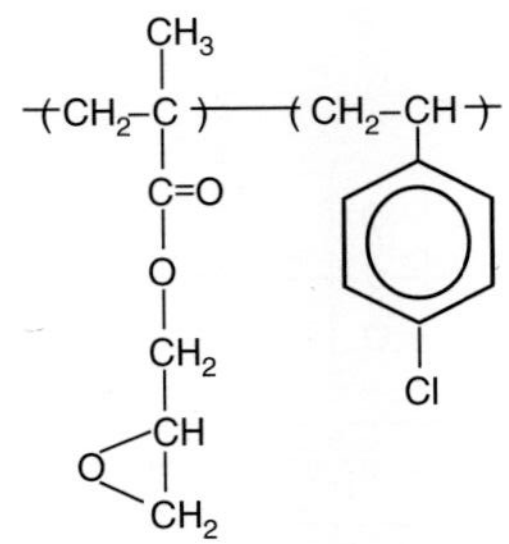	9

(*Continued*)

Table 3.1. (Continued)

Polymer title	Monomer unit	References
Poly(butane-1-sulfone)	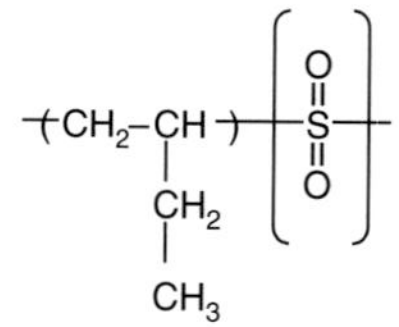	9
Novolac	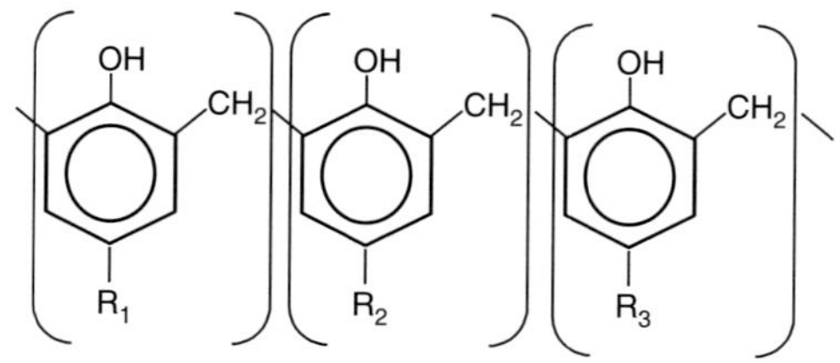	9
Polycarbosilane (PCS)	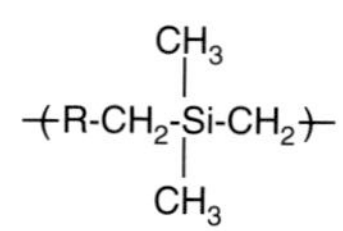	12

Polyphenylquinoxaline (PPQ)	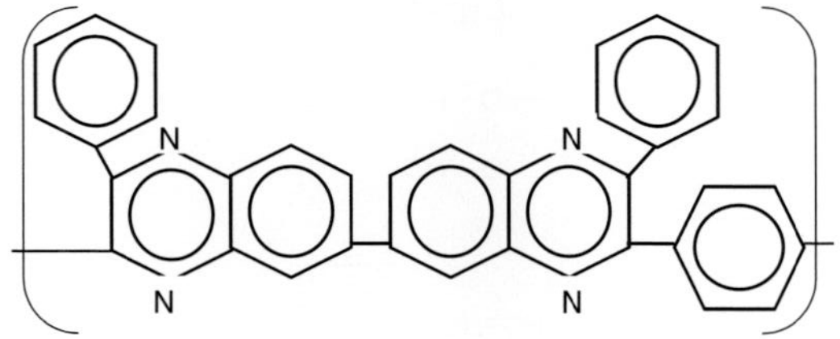	18
Poly-p-chlorstyrene	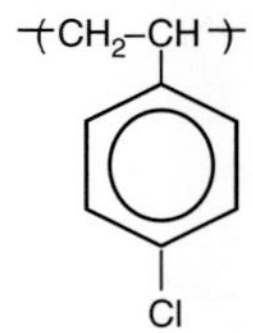	31
Polyoxymethylene	$-(CH_2-O)-$	32

(*Continued*)

Table 3.1. (Continued)

Polymer title	Monomer unit	References
Cellulose	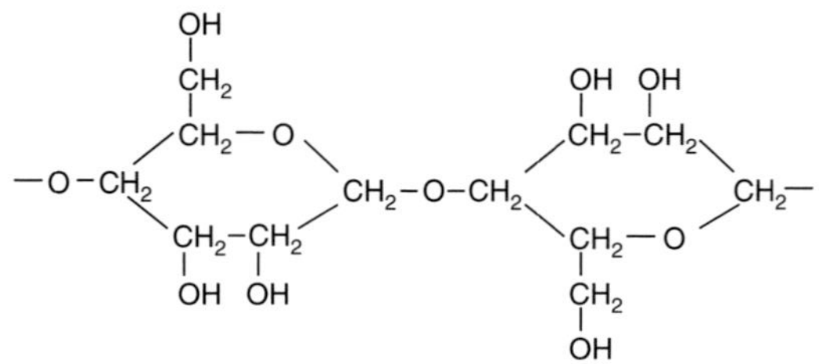	40
Polyphenylvinylene	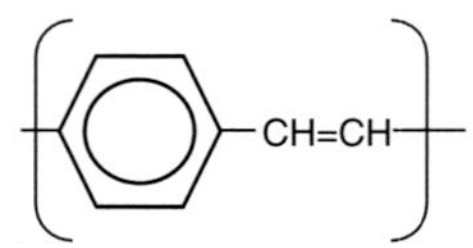	54
Poly(L-lactic acid) and co-polymers (PLLA, PLA, PLGA)	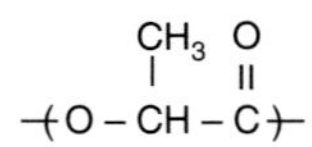	67, 117
Polyacrylonitrile and co-polymers	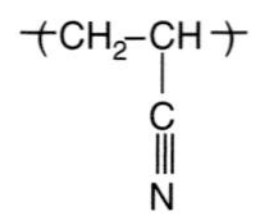	71

Poly(aryl ether ether ketone) (PEEK)	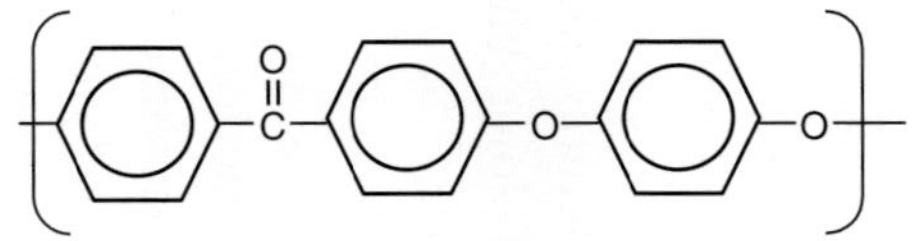	81
Polyacetylene	$-(CH{\equiv}CH)-$	101
Ethylene-propylene rubber (EPDM)	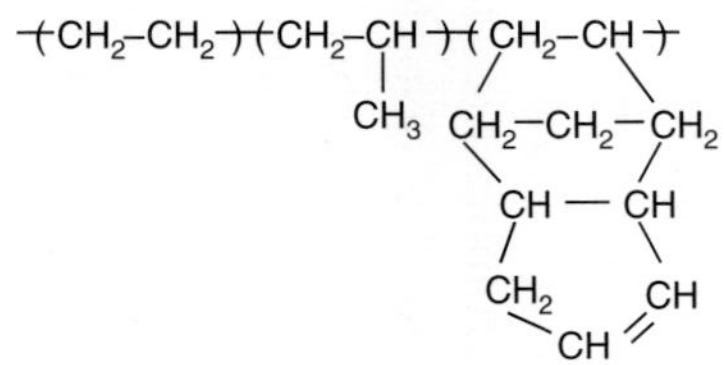	107
Poly(ethylene-tetrafluorethylene) (ETFE)	$-(CH_2\text{-}CH_2)(CF_2\text{-}CF_2)-$	107

Polymers are characterized by strong intramolecular interaction between atoms in the macromolecule, and weak intermolecular interaction between neighboring macromolecules. The intramolecular interactions in macromolecules are usually provided by covalent bonds. In some kinds of polymers intramolecular interactions are provided by ionic and donor/acceptor bonds. Polymer macromolecules have relatively high mobility at room temperature; they can change conformation and diffuse between neighbors. The macromolecular mobility is characterized by the glass-transition temperature (T_g), at which a polymer transforms from solid (glass state) to viscous liquid.

Polymer macromolecules form structures like clews or globules. Due to the disordering of long chains, polymers have mostly amorphous structures. Some polymers have partly ordered structures, while others have crystalline structures. For example, polyethylene has amorphous and crystalline fractions. A ratio of these fractions depends on the amount of structure defects in macromolecules. The crystalline fraction in low density polyethylene (LDPE) is about 40% and in ultra-high molecular weight polyethylene (UHMWPE) it is about 85–90%.

Industrial polymers contain additives: stabilizers, antioxidants, vulcanizing agents, plasticizers, fillers and other functional additives (sometimes up to 20–30 components) to provide required exploitation properties. Polymers can contain uncontrolled additives like minor products of synthesis reactions, oxidation and pollutions of production processes. Polymer surfaces can contain additional components like lubricants of the rolling process, antisticking powder or even fat traces from fingers after touch. For example, "fingerprints" are observed as carbon traces on Teflon surfaces after ion beam implantation with high fluence. Such additions must be removed from a polymer surface before modification. If complete removal is impossible, it must be taken into account at analysis after ion beam implantation.

A polymer surface can be cleaned with solvents. The choice of solvent is based on the compatibility of the polymer and solvent. If a polymer is immersed in solvent, the solvent penetrates into the surface layer of the polymer at cleaning. Also the solvent can bring low molecular weight components from bulk layers to the surface layer of polymers and different compositions of the surface layer before ion beam implantation can be obtained. In addition, the solvent (and other low molecular weight components of polymer) can evaporate into the vacuum chamber and contaminate the plasma. In particular, the effect of small molecule evaporation into the ion beam bar or plasma discharge is observed for highly plasticized rubbers, which usually contain 30–40% of low molecular weight plasticizers. Careful preparation of the polymer surface and knowledge of polymer bulk and surface contamination will help to achieve correct results at ion beam implantation.

A polymer macromolecule is stable when there is no external destruction influence. Under external influence (for example, UV light or ion beam) some bonds are dangled and a macromolecule is transformed into a free radical state. Such free radicals are very active and cause a number of chemical reactions in polymers. Here, for example, we consider free radical reactions in polyethylene, which was mostly investigated under radiation [119, 120]. Two kinds of radicals can be generated in the initial polyethylene macromolecule: the end radical appeared after the C—C bond break and the middle radical of the macromolecule appeared at the C—H bond break. The following reactions correspond to the generation of these two kinds of free radicals:

$$-\!\!\left(CH_2-CH_2-\dot{C}H-CH_2-CH_2 \right)\!\!- \;\longrightarrow\; -\!\!\left(CH_2-\dot{C}H-CH_2-CH_2-CH_2 \right)\!\!- \qquad (3.1)$$

$$\begin{matrix} -\!\left(CH_2-CH_2-\dot{C}H-CH_2-CH_2 \right)\!- & & -\!\left(CH_2-CH_2-CH_2-CH_2-CH_2 \right)\!- \\ + & \longrightarrow & + \\ -\!\left(CH_2-CH_2-CH_2-CH_2-CH_2 \right)\!- & & -\!\left(CH_2-CH_2-\dot{C}H-CH_2-CH_2 \right)\!- \end{matrix} \qquad (3.2)$$

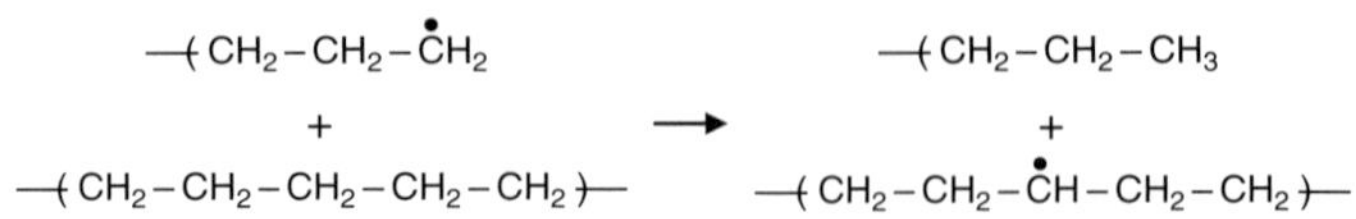

(3.3)

Reactions (3.1), (3.2) and (3.3) represent the transfer of the free valence bond from the excited macromolecule to its neighboring virgin macromolecule or along the macromolecule itself. Reactions (3.2) and (3.3) depend on the intermolecular environment and need strong intermolecular interaction between the two neighboring macromolecules. Consequently, the free radical moves into the deep layers of the polymer, which are not activated by penetrating external particles. In the case of ion beam implantation, such reactions can expand the modified layers more than the ion penetration depth layers.

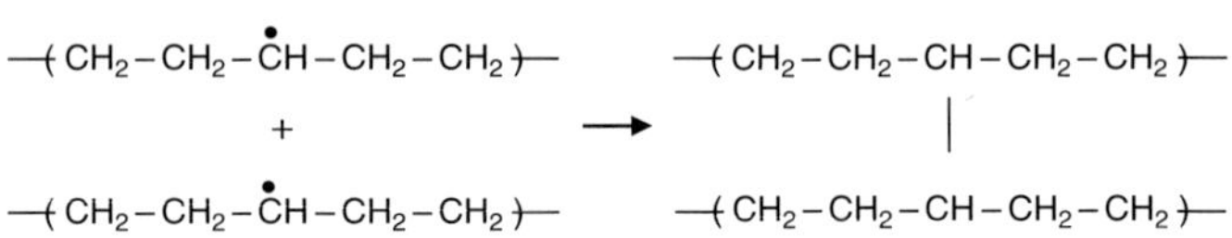

(3.4)

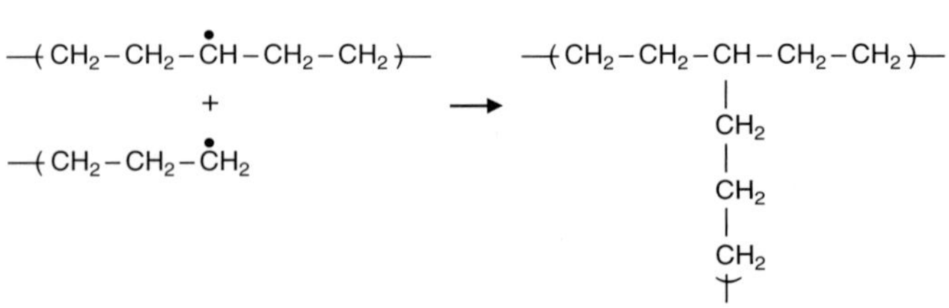

(3.5)

Reactions (3.4) and (3.5) generate crosslinks between macromolecules. These crosslinks are between the two centers of neighbouring macromolecules or form a T-bar between the centre of the first macromolecule. The reactions proceed when two free radicals of two different macromolecules meet.

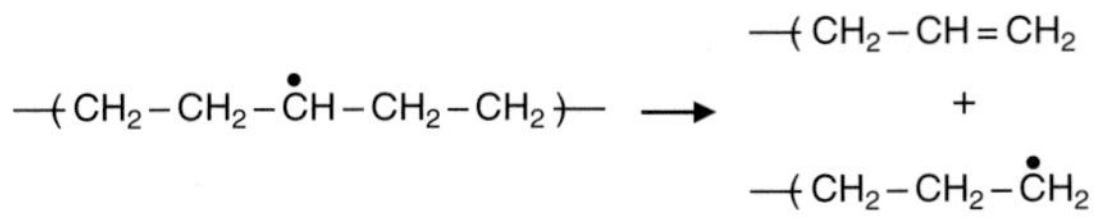

(3.6)

In reaction (3.6), the molecule backbone is broken and the carbon double-bond reappears. The free radical is conserved and can induce additional structural changes.

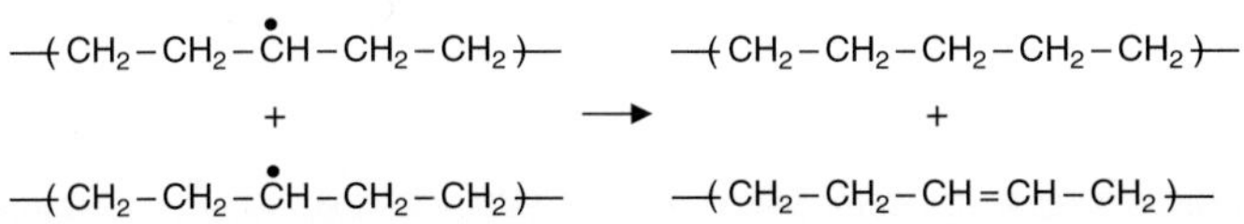

(3.7)

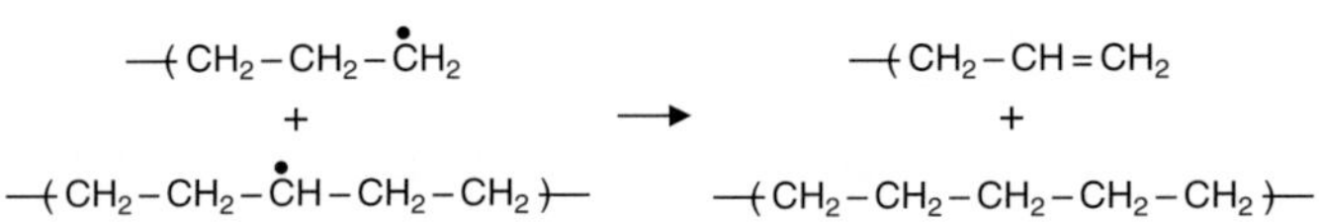

(3.8)

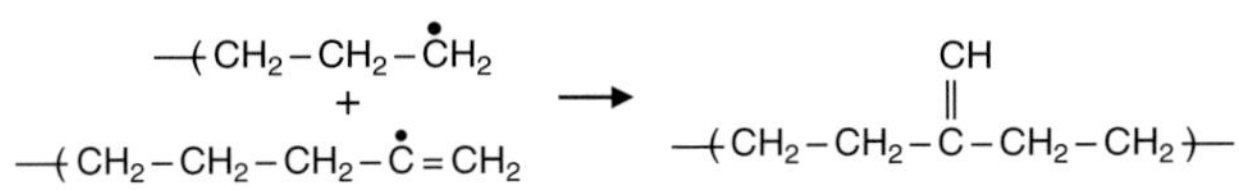

(3.9)

Reactions (3.7) and (3.9) occur in the presence of a neighboring macromolecule with a free radical which takes hydrogen from the neighboring macromolecule. As a result, the double carbon—carbon bond appears in the macromolecule. For the polyethylene macromolecule three kinds of double bonds are available: the vinylene group as a result of reaction (3.7); the vinyl group as a result of reaction (3.8); and the vinylidene group as a result of reaction (3.9).

The process of free radical transformation depends on neighboring macromolecules, phonon excitation and the electronic state of the macromolecule fragment. However, the free radical processes are mostly spontaneous and are described by probability functions.

In other polymers the free radical reactions are more complex. For example, the free radical may generate a depolymerization reaction, resulting in the decomposition of the macromolecules into several individual monomers. A classical example of the depolymerization reaction is the decomposition of butyl rubber to isobutylene monomers:

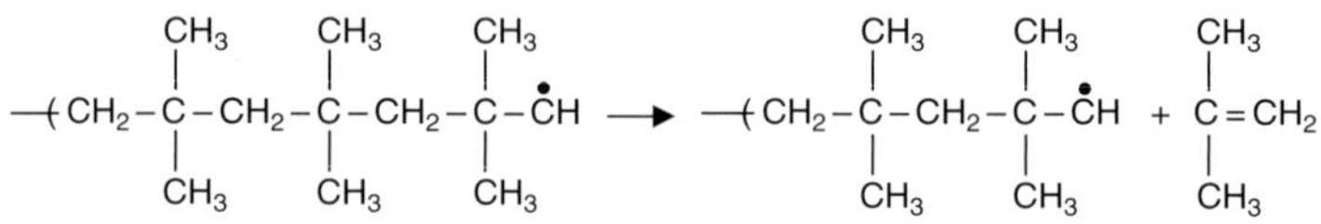

(3.10)

For some polymers the free radicals can initiate reactions that spontaneously release gaseous products, in the form of gaseous monomers or other molecules. For example, the free radical reaction in a carbonyl-containing polymer (for example, PMMA) causes the release of carbon monoxide gas.

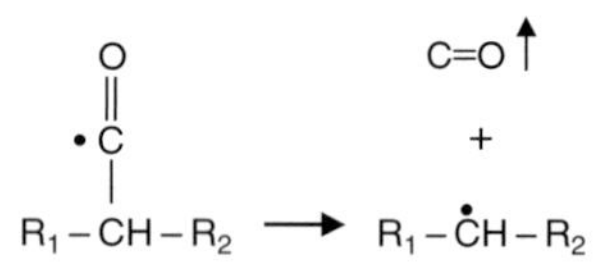

(3.11)

Related to depolymerization (scission) and crosslinking reactions at radiation influence, all polymers can be divided into two kinds: those where the scission reaction dominates and the polymer depolymerizes with time of irradiation treatment; and those where crosslinking reactions dominate and the polymer becomes crosslinked after exposure under irradiation. Examples of crosslinking polymers are polyethylene and polystyrene; examples of scission polymers are polyisobutylene and polymethylmethacrilate (PMMA).

In comparison with UV light, γ-irradiation and X-rays, ion beam implantation generates specific kinds of radicals. When a high energy ion collides with a macromolecule, the atom (or macromolecule fragment) leaves its position and moves at high speed away from the parent macromolecule. After a number of collisions and loss of kinetic energy the atom is stopped between polymer macromolecules. At this time the atom has dangled bonds. For example, if polyethylene is treated, carbon atoms with four free valence electrons each and hydrogen with one free valence electron appear at a certain distance from the initial track of the penetrating ion. Such recoiled atoms are extremely active radicals and they cause specific reactions.

Hydrogen atoms can join to the dangled bond of the macromolecule, to an unsaturated carbon—carbon bond or to another hydrogen atom with the formation of a hydrogen molecule.

Recoiled carbon atoms penetrating into hydrocarbon polymers can react with virgin macromolecules by the following scheme:

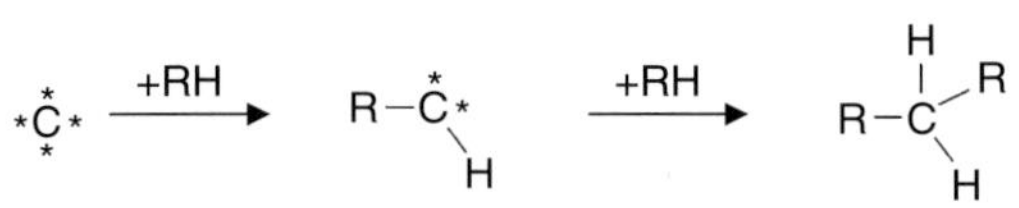

(3.12)

for example, with polyethylene macromolecules to give a crosslink:

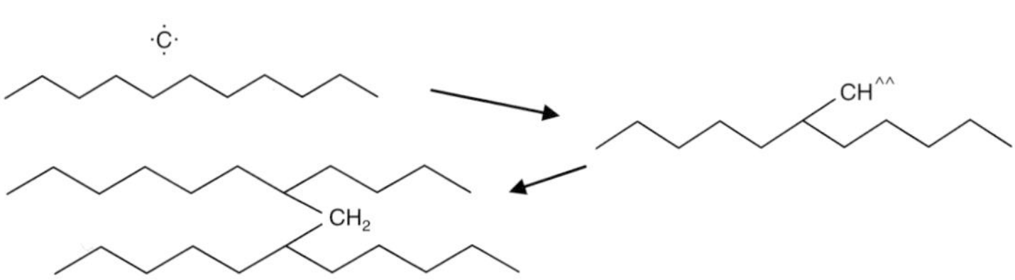

(3.13)

Such reactions are common at low fluence of ion beam implantation, when the concentration of carbon recoil atoms is low and the concentration of unchanged virgin macromolecules is high. As fluence increases, the amount of carbon recoil atoms increases and the recoiled carbon atom has a high probability of meeting a free radical and joining to it.

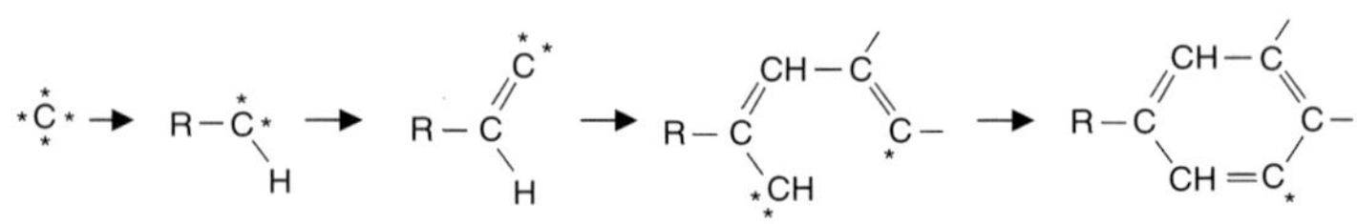

(3.14)

The following ion beam implantation generates an appearance of carbon clusters with irregular structure. A combination of unsaturated double and triple bonds, aromatic six and fifth atom rings, and conjugated structures are expected:

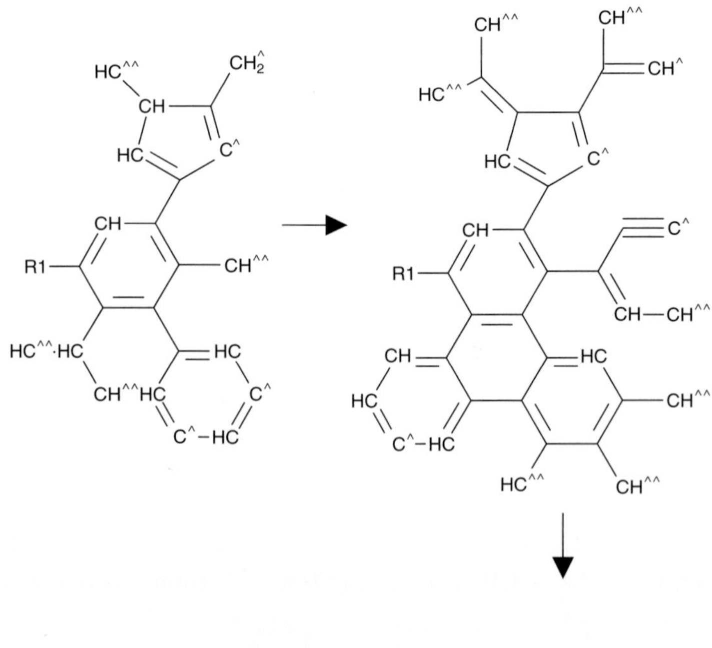

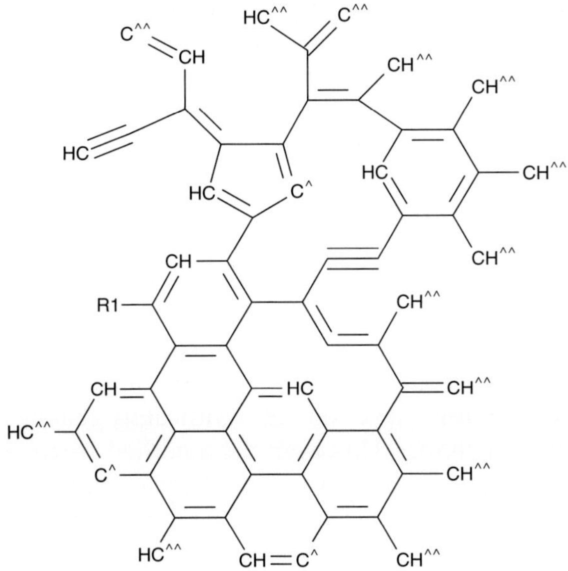

(3.15)

At such high fluence of implantation, the modified layer does not contain polymer macromolecules. This structure is characterized as pyrolytic carbon with short fragments of initial macromolecules. The size of the conjugated structures such as a length of $-(C{=}C)_n-$ polyene structures or a number of conjugated aromatic rings grows with fluence increase. For example:

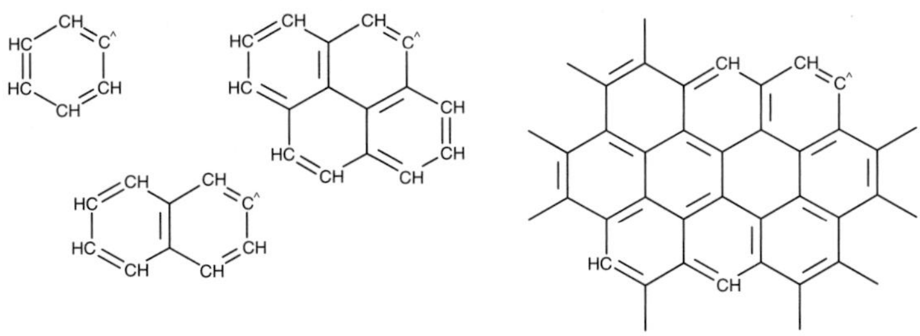

(3.16)

At high fluence such carbon clusters become sufficiently broad and fill the polymer surface. The penetrating ions collide the carbon clusters. At this fluence the ion beam implantation into the polymer finishes and an ion beam implantation into the carbon structure starts.

After completion of ion beam implantation the modified layer of the polymer is unstable. The free radicals survive for a long time after initiation. Despite high speed reactions at the first period of time, the presence and activity of free radicals is observed from a few days to even months after treatment. In other words, the modified layer is "living". If a modified polymer is used after ion beam implantation in high vacuum (for example, in an ion implanter chamber or in free space environment), the free radical reactions continuing in the modified layer do not significantly change the structure of the polymer. However, the modified polymer is usually taken away from the vacuum chamber and used in air or in another active environment. Due to the presence of oxygen, the reactions of free radicals with oxygen proceed and change significantly the structure of polymer on exposure to air. The reaction of a free radical with oxygen in air starts with following scheme:

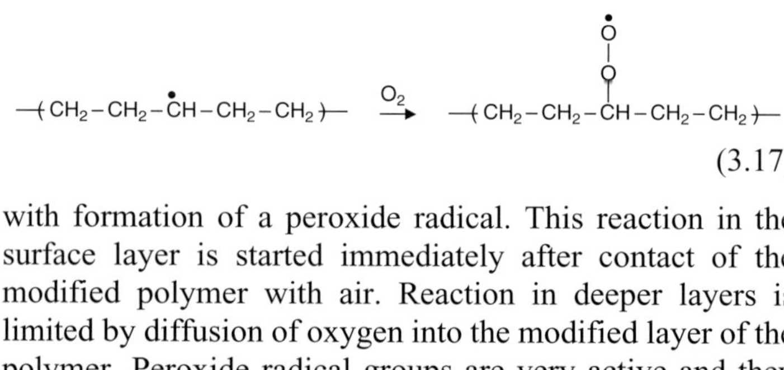

(3.17)

with formation of a peroxide radical. This reaction in the surface layer is started immediately after contact of the modified polymer with air. Reaction in deeper layers is limited by diffusion of oxygen into the modified layer of the polymer. Peroxide radical groups are very active and they can react with free alkyl radicals to give peroxide crosslinks:

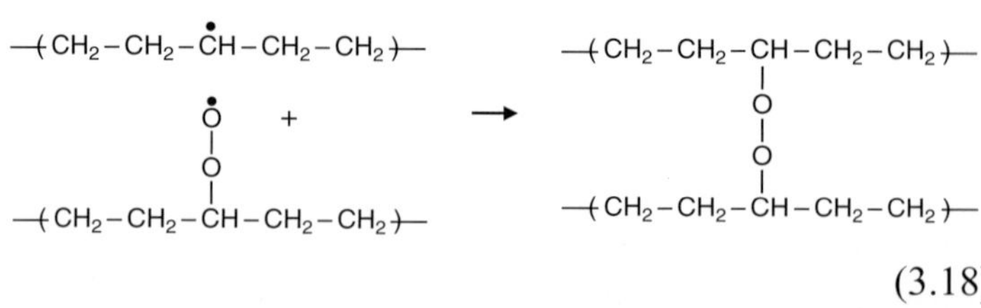

(3.18)

The peroxide radical can react with a neighboring hydrocarbon molecule, easily take in hydrogen and form a hydroperoxide group:

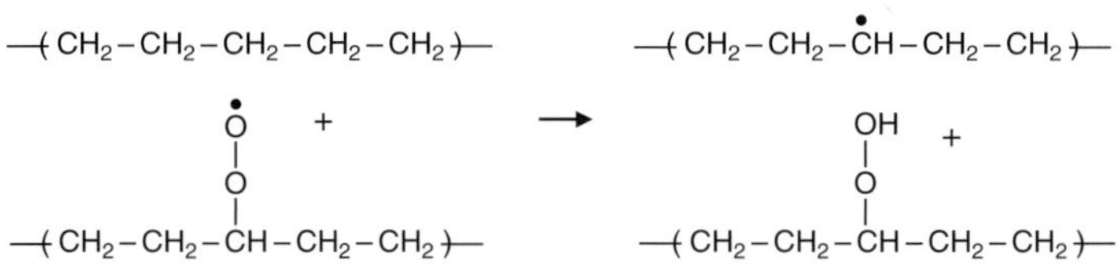

(3.19)

The peroxide radical can also take hydrogen from a neighboring group of the same molecule and form a hydrocarbon free radical. For the polyethylene macromolecule, this reaction can occur in the following way:

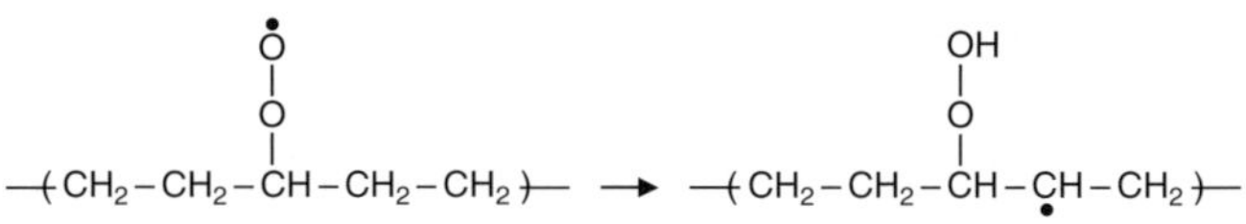

(3.20)

The peroxide group is more stable than the free radicals and it can exist in the modified layer of the polymer for some hours and even days. Over time the peroxide group can degrade by the following scheme:

$$ROOH \rightarrow R^* + {}^*OOH \quad (3.21)$$

$$ROOH \rightarrow RO^* + {}^*OH \quad (3.22)$$

$$ROOH \rightarrow ROO^* + {}^*H \quad (3.23)$$

or react with free radicals

$$ROOH + RO^* \rightarrow ROO^* + ROH \quad (3.24)$$

$$ROOH + R^* \rightarrow ROO^* + RH \quad (3.25)$$

Reactions (3.21), (3.23) and (3.25) are reversible and do not give new products. Reaction (3.24) gives a stable hydroxyl group. Reaction (3.22) gives an alkoxyl radical which is unstable and can be transformed into stable groups by the following schemes:

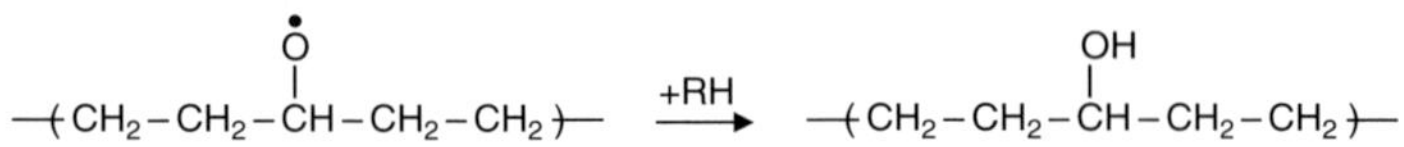

(3.26)

of the hydroxyl group,

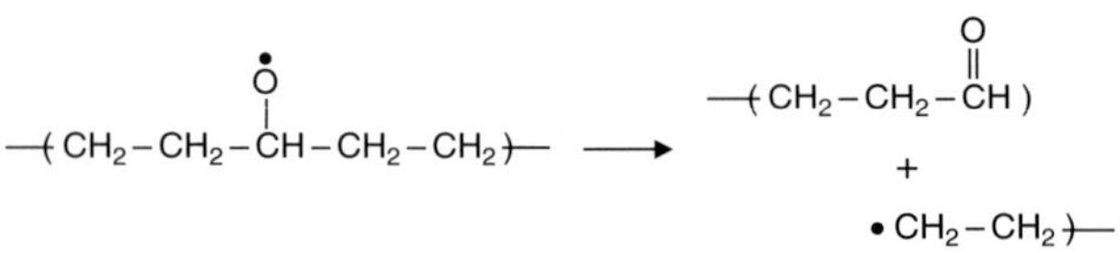

(3.27)

of the aldehyde group,

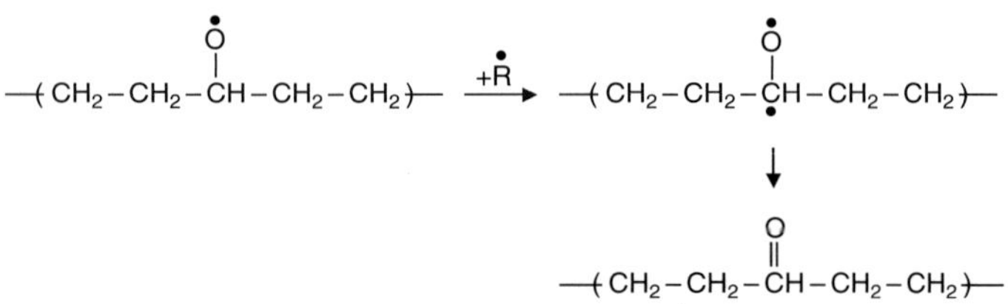

(3.28)

of the ketone group,

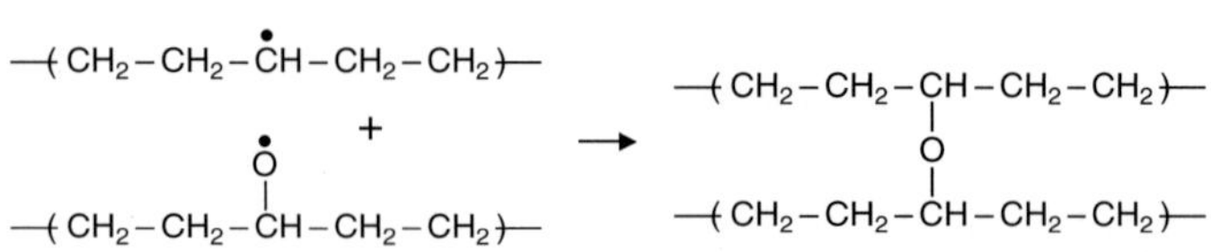

(3.29)

and of the ether group.

The two interacted alkoxyl radicals can give directly ketone and hydroxyl groups due to the reaction of disproportionation.

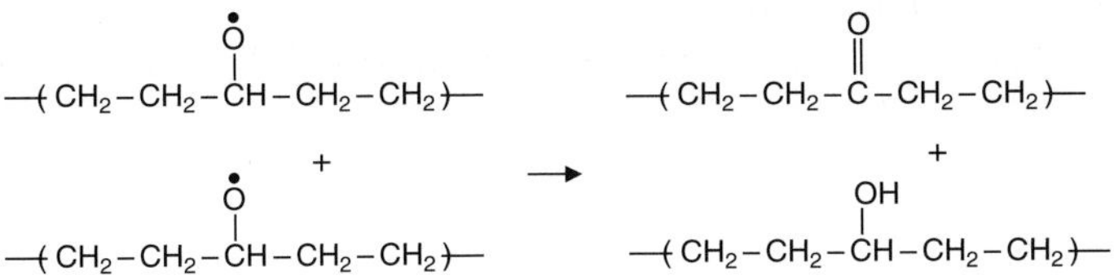

(3.30)

The aldehyde group can react with an oxidizing agent and form a carboxyl group.

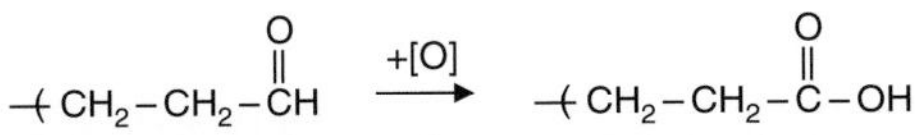

(3.31)

Such reactions proceed both in the modified layer and in the deeper layer, in which free radicals move under reactions of free valence transfer.

At high fluence of implantation, the oxidation of a highly carbonized layer also occurs. After the first stage of free radical transformation in the modified layer the cluster structure of condensed aromatic cycles is formed. Free radicals reside predominantly on the edges of the cycles. Therefore, the reactions with atmospheric oxygen proceed mostly on the edges of the cycles. An example of the structure of the carbon cluster after oxidation and formation of hydroxyl, aldehyde, carboxyl, ether and peroxide groups can be visualized as follows:

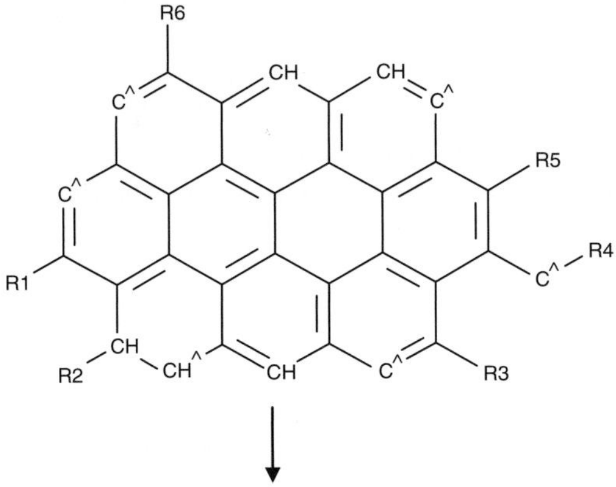

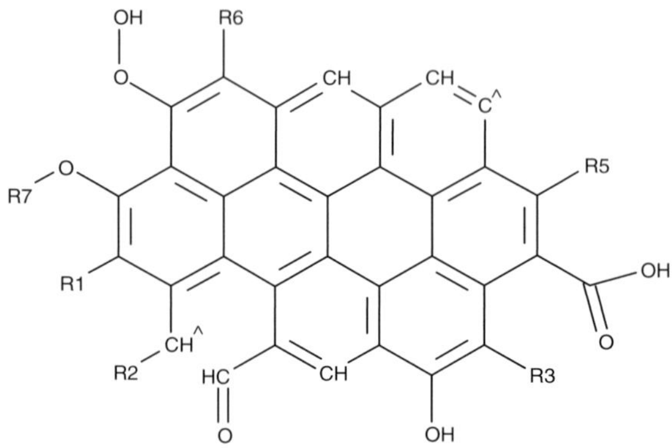

(3.32)

The modified surface contains free radicals that are not only active to atmospheric oxygen but also to a wide number of substances with different kinds of reactionable groups. One of the examples is the alkene group, which is active to be jointed with free radicals by the following reaction:

$$R_1-CH=CH-R_2+\dot{R}_3 \rightarrow R_1-\underset{\bullet}{C}H-\overset{R_3}{\overset{|}{C}H}-R_2 \qquad (3.33)$$

By this reaction any functional compounds could be attached to the modified polymer surface. For example, acrylic substances can react with free radicals:

$$HO-\overset{O}{\overset{\|}{C}}-CH=CH_2+\dot{R}_3 \rightarrow HO-\overset{O}{\overset{\|}{C}}-\underset{\bullet}{C}H-\overset{R_3}{\overset{|}{C}H_2} \qquad (3.34)$$

and form a polyacrylic acid layer on the modified polymer:

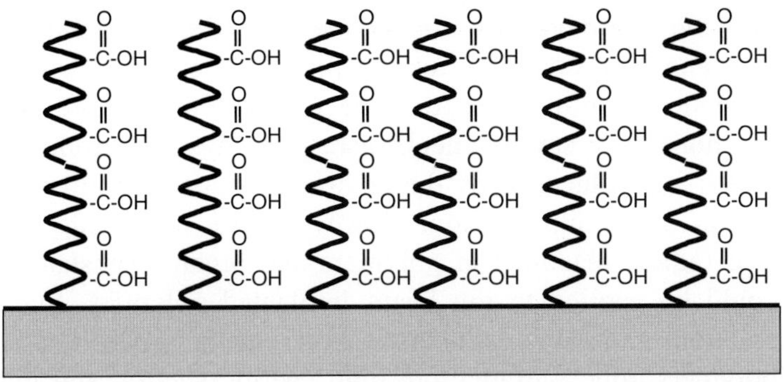

(3.35)

Or the amine layer could be created by the reaction of Allylamine:

$$H_2N-\overset{\overset{\displaystyle O}{\|}}{C}-CH=CH_2+\dot{R}_3 \rightarrow H_2N-\overset{\overset{\displaystyle O}{\|}}{C}-\underset{\bullet}{C}H-\overset{\overset{\displaystyle R_3}{|}}{C}H_2 \qquad (3.36)$$

After reactions with active substances, the surface has renewed chemical activity and can other components may attach. For example, the surface after ion beam and acrylic acid becomes active to base substances:

$$R_1-\overset{\overset{\displaystyle O}{\|}}{C}-OH+H_2N-R_2 \rightarrow R_1-\overset{\overset{\displaystyle O}{\|}}{C}-NH-R_2+H_2O \qquad (3.37)$$

where R_1 can be linker molecule, polyamine glue or protein molecule.

Useful examples with experimental data of chemically attached substances will be considered in following chapters.

The above chemical reactions in polymers after ion beam implantation have multipath characteristics. The propagation paths, rates and products of the reaction are difficult to predict in terms of polymer type, ion beam parameters and post-history of the polymer. However, some common principles of the reactions can be overviewed. The reactions of free radicals begin immediately after ion penetration into the polymer. Reaction kinetics are difficult to measure and investigate because of their high speed and their dependence on distance from the surface. Furthermore, the reactions have several mechanism pathways and products, which are influenced by the environment after termination of ion beam implantation. However, basic knowledge of free radical reactions in polymers and organic substances could be applied to ion beam implanted polymers. Such reactions and their products are experimentally observed in polymers after γ-irradiation, high energy electron and ion beams, X-ray irradiation and UV light. The rate of these reactions depend on the mobility of the free radicals, macromolecules and their activity. Fundamentally the reaction kinetics are described by Arrhenius law:

$$\frac{\partial[R_i\bullet]}{\partial t} = k[R_i\bullet][R_jH] \quad (3.1)$$

where $[R_i\bullet]$ is the concentration of ith free radical, $[R_jH]$ is the concentration of unchanged macromolecular fragment, k is the reaction rate which depends on temperature (T):

$$k(T) = k_0 e^{-\frac{E}{KT}} \quad (3.2)$$

Reaction pathways and rates depend on the local structure of the polymer (crystalline or amorphous phases), the neighbouring atoms/macromolecules and the conformation of the macromolecules. Reaction kinetics can be described using several different reaction rates which correspond to localised differences in structure. These reaction kinetics are called "polychronos kinetics" and are based on the $f(E)$ distribution function of active free radicals by activation energy (E) [120]. The concentration of active radicals can be expressed as an integral on different ensembles of radicals, each with different activation energy of the radical reactions:

$$n(t, T) = n_0 \int_{E_{\min}}^{E_{\max}} f(E) \cdot G[k(E), t] \cdot dE \qquad (3.3)$$

where $G[k(E),t]$ is a kinetic equation for ensembles with E activation energy and n_0 is the initial concentration of the radicals.

Experimentally the kinetics of the free radical reactions are observed by unfreezing the free radicals step-by-step at elevated temperatures (Fig. 3.1). The kinetic curve of the integral amount of free radicals with time after irradiation can be described by a number of exponential functions and pre-exponential coefficients. The character of free radical reactions is observed for a wide number of polymers, radical reactions and methods of free radical generations. Ion beam implantation is not an exception to this number.

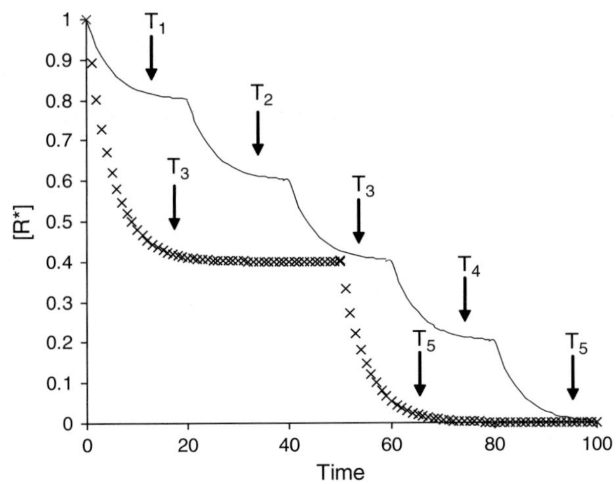

Fig. 3.1. Model curve of polychronos kinetics of unfreezing of free radicals at step heating of a polymer. Signed curve corresponds to two steps; continuous curve corresponds to five steps at the same temperature range.

However, ion beam implantation is an exception when considering the high concentration of active free radicals at high local temperature at the same time, in small volume. Calculated phonon temperature in the track region of a penetrating ion can reach 10^4 K which is significantly higher than any temperature for polymer surviving. Therefore, the first period of time after ion propagation, the free radical reactions proceed in limited volume and at extremely high temperatures. When the temperature spike relaxes, the kinetics of reaction become sensitive to the temperature of the sample holder, the environment and the ion beam current density. Even in a low current density regime without overheating the sample, the temperature near the ion track remains higher than the temperature of the bulk sample. If ion current is dense enough and a second ion penetrates the region of the polymer close to the first ion track, the temperature spikes from both ions can be overlapped. Therefore, the free radical reactions in this region of the polymer proceed at a higher temperature but at a lower current density. This effect changes the path and rates of free radical reactions at a high current density regime of the ion beam implantation. This holds even if the temperature increase of the bulk material and the surface layer is small. Experimentally these effects are observed by comparing continuous and pulse ion beam implantation regimes with equal average current density. The pulse regime has higher ion density during the pulse than the continuous regime. Therefore, it is expected that the products of free radical reactions using continuous and pulse implantation regimes are different, even though polymer temperature is the same during both regimes of ion beam implantation.

Despite the high rate of free radical reactions the concentration of free radicals does not decrease to zero after storing the implanted polymer for any length of time. The free radicals remain in irradiated polyethylene for some months and in irradiated polytetrafluorethylene for some years. This prolonged activity is due to the long life time of specific types of free radicals and the relatively stable middle products of free radical reactions (peroxide groups, for example). In addition, free radicals can be trapped at the edges of aromatic structure where the unpaired electron of free radical is stabilized by presence of π-electrons in the conjugated aromatic structure.

After ion beam implantation, the active components of the environment can react with residual free radicals in the polymer surface layer. The kinetic equations that describe the reactions include the concentration of the active compound. For example, the reactions with atmospheric oxygen depend on the concentration of oxygen in gas media over the polymer surface:

$$\frac{\partial[R_iOO\bullet]}{\partial t} = k[R_i\bullet][O_2] \tag{3.4}$$

where $[R_i\bullet]$ is the concentration of ith residual free radical, $[R_jOO\bullet]$ is the concentration of the product of oxidation and $[O_2]$ is the concentration of oxygen in the polymer layer.

If the concentration of oxygen is low but the polymer is stored under vacuum for long enough, oxygen-containing groups can appear in the irradiated polymer even if the polymer does not come into contact with the atmosphere after ion beam implantation.

The same effect is observed with the presence of residual gases in the vacuum chamber (oil from vacuum pumps, grease or destructive products from previous polymers). The polymer surface becomes very attractive to any reactionable gaseous substances after ion beam implantation and even low concentrations of residual gases or vapors are enough to react with free radicals in the modified surface layer.

Despite the high concentration of initial free radicals in the modified polymer, the relative concentration of active free radicals after removing the sample from the vacuum chamber to the atmosphere is quite low. If the concentration of free radicals is not enough to reach the required activity of the polymer surface, the concentration can be increased using an interrupted regime of treatment: the polymer after ion beam implantation is removed to external active media and after activation the ion beam treatment is repeated with activation to reach the required properties of the polymer surface.

For example, the periodic influence of atmospheric oxygen after treatment could be used to dramatically increase the concentration of oxygen-containing groups in the surface layer of the modified polymer. The modification process may be considered in terms of the simple kinetics of defect collections under irradiation. The activation of the surface by irradiation on active centers of concentration (R) has asymptotic dependence on treatment time (t_1):

$$R = R_0 \cdot (1 - \exp(-k_1 t_1)) \tag{3.5}$$

where k_1 is the rate of activation and R_0 is the maximal concentration of active centers (free radicals). After treatment the active centers recombine with time (t_2). The concentration of the centers (R') decreases by:

$$R' = R \cdot \exp(-k_2 t_2) \tag{3.6}$$

where k_1 is a rate of recombination. If the recombination occurs in the presence of an active environment (for example, oxygen), the concentration of the active centers decreases by:

$$dR' = -k_2 \cdot R' \cdot dt_2 - k_3 \cdot R' \cdot dt_2 \tag{3.7}$$

where k_3 is a rate of associations with active environmental molecules. The concentration of jointed molecules (C) corresponds to:

$$dC = k_3 \cdot \alpha \cdot R' \cdot dt_2 \tag{3.8}$$

where a is the yield of chemical reaction with active environmental molecules.

After integration

$$R' = R \cdot \exp\{-(k_2 + k_3) \cdot t_2\} \tag{3.9}$$

$$C = \frac{k_3 \cdot \alpha}{k_2 + k_3} \cdot R \cdot [1 - \exp\{-(k_2 + k_3) \cdot t_2\}] + C_0 \tag{3.10}$$

If the treatment is repeated after exposure of the modified polymer to an active environment, the concentration of the jointed molecules could be expressed as (3.10):

First treatment

$$\begin{aligned} R_1 &= R_0 \cdot \{1 - \exp(-k_1 t_1)\} \\ R_1' &= R_1 \cdot \exp(-(k_2 + k_3)t_2) \\ &= R_0 \cdot \exp(-(k_2 + k_3)t_2) \cdot \{1 - \exp(-k_1 t_1)\} \\ C_1 &= \tfrac{k_3 \cdot \alpha \cdot R_1}{k_2 + k_3} \cdot \{1 - \exp(-(k_2 + k_3)t_2)\} + C_0 \end{aligned}$$

Second treatment

$$\begin{aligned} R_2 &= R_0 \cdot \{1 - \exp(-k_1 t_1)\} + R_1' \cdot \exp(-k_1 t_1) \\ R_2' &= R_2 \cdot \exp(-(k_2 + k_3)t_2) \\ &= R_0 \cdot \exp(-(k_2 + k_3)t_2) \cdot \{1 - \exp(-k_1 t_1)\} \\ &\quad + R_1' \cdot \exp(-(k_2 + k_3)t_2) \cdot \exp(-k_1 t_1) \\ C_2 &= \tfrac{k_3 \cdot \alpha \cdot R_2}{k_2 + k_3} \cdot \{1 - \exp(-(k_2 + k_3)t_2)\} + C_1 \end{aligned}$$

....

(n − 1) treatment

$$\begin{aligned} R_{n-1} &= R_0 \cdot \{1 - \exp(-k_1 t_1)\} + R_{n-2}' \cdot \exp(-k_1 t_1) \\ R_{n-1}' &= R_{n-1} \cdot \exp(-(k_2 + k_3)t_2) \\ C_{n-1} &= \tfrac{k_3 \cdot \alpha \cdot R_{n-1}}{k_2 + k_3} \cdot \{1 - \exp(-(k_2 + k_3)t_2)\} + C_{n-2} \end{aligned}$$

n(last) treatment

$$\begin{aligned} R_n &= R_0 \cdot \{1 - \exp(-k_1 t_1)\} + R_{n-1}' \cdot \exp(-k_1 t_1) \\ C_n &= \tfrac{k_3 \cdot \alpha \cdot R_n}{k_2 + k_3} \cdot \{1 - \exp(-(k_2 + k_3)t_2)\} + C_{n-1} \end{aligned} \tag{3.11}$$

If the measurement of the jointed molecules on an n-time treated polymer is done after a sufficiently long period of time ($t_2 \to \infty$), the concentration of the jointed molecules will be:

$$C_n = \frac{k_3 \cdot \alpha \cdot R_0}{k_2 + k_3} \cdot \{1 - \exp(-k_1 t_1)\} \times \left[\sum_{i=0}^{n-1} (n-i) \cdot \exp\{-ik_1 t_1 - i(k_2 + k_3)t_2\} - \sum_{i=1}^{n-1} (n-i) \cdot \exp\{-(i-1)k_1 t_1 - i(k_2 + k_3)t_2\} \right] \tag{3.12}$$

The interrupted regime of five-time plasma treatment was applied for EPDM rubber. The concentration of oxygen-containing groups was measured by FTIR ATR spectra

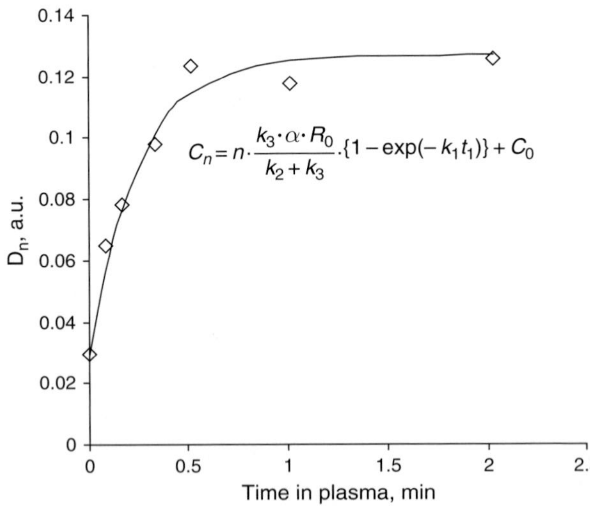

Fig. 3.2. Absorbance of carbonyl line in FTIR ATR spectra of EPDM at time of plasma treatment.

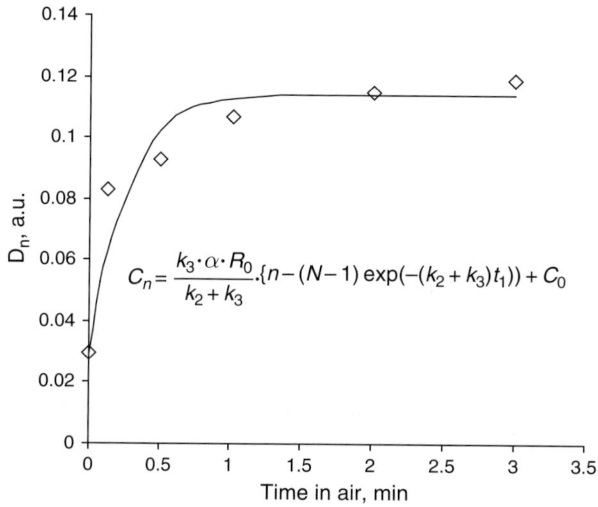

Fig. 3.3. Absorbance of carbonyl line in FTIR ATR spectra of EPDM at time of exposure in air after plasma treatment.

as a function of the exposure time to the plasma discharge, and the exposure time to the atmosphere between treatments (Figs 3.2 and 3.3). The curves show that the concentration of oxygen-containing groups depends on the time in plasma discharge and on the time of exposure to air after treatment as given by the asymptotical law in equations (3.5) and (3.8). These curves were used to calculate the coefficients in equation (3.12), which in turn were used for calculation of the interrupted regime of multiple treatment (Fig. 3.4). Good agreement of the experimental and theoretical data was observed. The deviation of experimental results at high concentration of oxygen-containing groups is caused by the significant etching effect of the oxidized polymer layer which means that the maximal value of the oxygen-containing groups can be jointed to the polymer surface. This maximal concentration is ten times higher than the concentration that could be distinguished by a continuous regime of treatment.

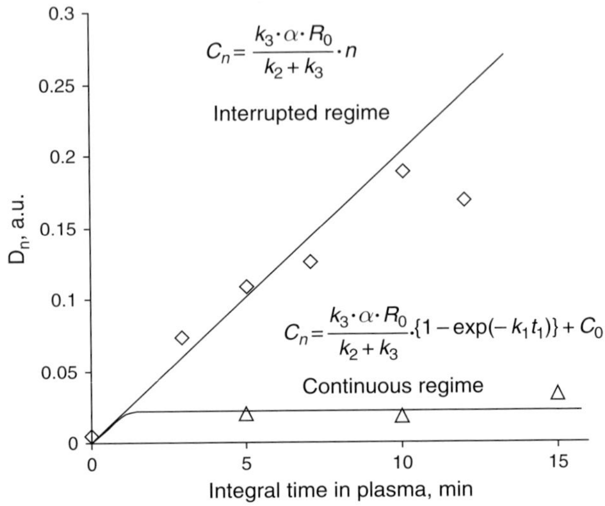

Fig. 3.4. Absorbance of carbonyl line in FTIR ATR spectra of EPDM at integral time of plasma treatment for continuous and interrupted regimes (experimental points and theoretical curves).

References

1. M. Manso, A. Valsesia, M. Lejeune, D. Gilliland, G. Ceccone, F. Rossi, Tailoring surface properties of biomedical polymers by implantation of Ar and He ions, *Acta Biomaterialia* 1 (2005) 431–440.
2. W. Hong, H.-J. Woo, H.-W. Choi, Y.-S. Kim, G. Kim, Optical property modification of PMMA by ion-beam implantation, *Appl. Surf. Sci.* 169–170 (2001) 428–432.
3. J. Zhang, X. Yu, H. Li, X. Liu, Surface modification of polytetrafluoroethylene by nitrogen ion implantation, *Appl. Surf. Sci.* 185 (2002) 255–261.
4. J. Zhang, X. Zhang, H. Zhou, Effect of aging on surface chemical bonds of PTFE irradiated by low energy Ti ion, *Appl. Surf. Sci.* 205 (2003) 343–352.
5. M.M. Silvan, A. Valsesia, D. Gilliland, G. Ceccone, F. Rossi, An evaluation of poly(ethylene-glycol) films stabilized by plasma and ion beam methods, *Appl. Surf. Sci.* 235 (2004) 119–125.

6. J. Wang, N. Huang, P. Yang, Y.X. Leng, H. Sun, Z.Y. Liu, P.K. Chu, The effects of amorphous carbon films deposited on polyethylene terephthalate on bacterial adhesion, *Biomaterials* 25 (2004) 3163–3170.
7. W. Zhang, P.K. Chu, J. Ji, Y. Zhang, X. Liu, R.K.Y. Fu, P.C.T. Ha, Q. Yan, Plasma surface modification of polyvinylchloride for improvement of antibacterial properties, *Biomaterials* 27 (2006) 44–51.
8. K. Kurotobi, M. Kaibara, Y. Suzuki, M. Iwaki, H. Nakajima, S. Kaneko, Ion implantation into collagen-coated surfaces for the development of small diameter artificial grafts, *Coll. Surf. B: Biointerfaces* 19 (2000) 227–235.
9. T.M. Hall, A. Wagner, L.F. Thompson, Ion beam exposure characteristics of resists: experimental results, *J. Appl. Phys.* 53(6) (1982) 3997–4010.
10. E. Yap, D.G. McCulloch, D.R. McKenzie, M.V. Swain, L.S. Wielunski, R.A. Clissold, Modification of the mechanical and optical properties of a polycarbonate by 50 keV Ar+ and H+ ion implantation, *J. Appl. Phys.* 83(6) (1998) 3404–3412.
11. J.S. Chen, Z. Sun, P.S. Guo, Z.B. Zhang, D.Z. Zhu, H.J. Xu, Effect of ion implantation on surface energy of ultrahigh molecular weight polyethylene, *J. Appl. Phys.* 93(9) (2003) 5103–5108.
12. E.Z. Kurmaev, R.P. Winarski, J.-C. Pivin, D.L. Ederer, S.N. Shamin, A. Moewes, K. Endo, T. Ida, G.S. Chang, C.N. Whang, Chemical reactions in polymers induced by ion beam mixing: fluorescence X-ray measurements, *J. Elect. Spectros. Rel. Phen.* 110–111 (2000) 87–103.
13. J. Zhang, Q. Wua, X. Yu, P. Zha, H. Li, Effect of aging on the morphology and wettability of polytetrafluoroethylene, *Mat. Lett.* 48 (2001) 362–368.
14. K.Y. Rhee, J.H. Choi, S.J. Park, Effect of 1 keV Ar+ irradiation on the residual strength of PE fiber-reinforced composites, *Mat. Sci. Eng. A* 395 (2005) 288–294.
15. C. Satriano, N. Spinella, M. Manso, A. Licciardello, F. Rossi, G. Marletta, Ion beam induced nanometric structure and oligopeptide adsorption on patterned polymer surfaces, *Mat. Sci. Eng. C* 23 (2003) 779–786.

16. W. He, D.B. Poker, K.E. Gonsalves, N. Batina, Micro/nano machining of polymeric substrates by ion beam techniques, *Micro. Eng.* 65 (2003) 153–161.
17. J. Davenas, G. Boiteux, X.L. Xu, E. Adem, Role of the modifications induced by ion beam irradiation in the optical and conducting properties of polyimide, *Nucl. Instr. Meth. Phys. Res.* B32 (1988) 136–141.
18. J. Davenas, X.L. Xu, G. Boiteux, D. Sage, Relations between structure and electric properties of ion irradiated polymers, *Nucl. Inst. Meth. Phys. Res.* B39 (1989) 754–763.
19. A. Licciadello, O. Puglisi, L. Calcagno, G. Foti, Crosslinking yield in particle beam irradiated polystyrene, *Nucl. Inst. Meth. Phys. Res.* B46 (1990) 338–341.
20. J. Davenas, P. Thevenard, G. Boiteux, M. Fallavier, X.L. Xu, Hydrogenated carbon layers produced by ion beam irradiation of PMMA and polystyrene films, *Nucl. Inst. Meth. Phys. Res.* B46 (1990) 317–323.
21. J.P. Biersack, R. Kallweit, Ion beam induced changes of the refractive index of PMMA, *Nucl. Inst. Meth. Phys. Res.* B46 (1990) 309–312.
22. G. Marletta, Chemical reactions and physical property modifications induced by keV ion beams in polymers, *Nucl. Inst. Meth. Phys. Res.* B46 (1990) 295–305.
23. G. Foti, R. Reitano, Structure and bonding in ion irradiated polystyrene, *Nucl. Inst. Meth. Phys. Res.* B46 (1990) 306–308.
24. G.W. Arnold, R.R. Rye, Ion beam analysis of the effects of radiation on the chemical etching of poly(tetrafluorethylene), *Nucl. Inst. Meth. Phys. Res.* B46 (1990) 330–333.
25. Y. Suzuki, C. Swapp, M. Kusakabe, Aging effects on wettability and structure of ion implanted silicone, *Nucl. Inst. Meth. Phys. Res.* B46 (1990) 354–357.
26. L. Calcagno, G. Foti, Ion irradiation of polymers, *Nucl. Inst. Meth. Phys. Res.* B59/60 (1991) 1153–1158.
27. J. Davenas, P. Thevenard, Electronic structure characterization of ion beam modified polyimide by optical absorption and reflection, *Nucl. Inst. Meth. Phys. Res.* B59/60 (1991) 1249–1252.

28. L. Calcagno, G. Compagnini, G. Foti, Structural modification of polymer film by ion irradiation, *Nucl. Inst. Meth. Phys. Res.* B62 (1992) 413–422.
29. M.B. Lewis, E.H. Lee, Chemical G-value of ion irradiated polymers, *Nucl. Inst. Meth. Phys. Res.* B62 (1992) 341–348.
30. R.M. Papaleo, M.A. de Araujo, R.P. Livi, Study of the ion beam induced amorphisation, bond breaking and optical gap change processes in PET, *Nucl. Inst. Meth. Phys. Res.* B62 (1992) 442–446.
31. A.V. Leontyev, E.F. Ostretsov, V.V. Grigoryev, F.F. Komarov, Analytical methods for investigation of ion-implanted polymer layers, *Nucl. Inst. Meth. Phys. Res.* B62 (1992) 438–441.
32. R.M. Mahfouz, M. Sauer, S.T. Atwa, R.I. Kaiser, K. Roessler, Interaction of MeV ions and VUV photons with polymers, *Nucl. Inst. Meth. Phys. Res.* B62 (1992) 447–451.
33. Y. Suzuki, M. Kusakabe, J.-S. Lee, M. Kaibara, M. Iwaki, H. Sasabe, Endothelian cell adhesion to ion implanted polymers, *Nucl. Inst. Meth. Phys. Res.* B62 (1992) 142–147.
34. L. Dejun, Z. Jie, G. Hanqing, L. Mozhu, D. Fuqing, Z. Qiqing, Surface modification of medical polyurethane by silicon ion bombardment, *Nucl. Inst. Meth. Phys. Res.* B82 (1993) 57–62.
35. L. Calcagno, R. Percolla, G. Foti, Interdiffusion in polystyrene crosslinked by ion irradiation, *Nucl. Inst. Meth. Phys. Res.* B91 (1994) 426–430.
36. W.E. Wallace, T.T. Chiou, J.B. Rothman, R.J. Composto, Gas absorption during ion-irradiation of a polymer target, *Nucl. Inst. Meth. Phys. Res.* B103 (1995) 435–439.
37. F. Abel, V. Quillet, M. Schott, Degradation of polystyrene thin films under d, 4He and 12C irradiation studied by ion beam analysis: effects of energy loss, sample thickness and isotopic content, *Nucl. Inst. Meth. Phys. Res.* B105 (1995) 86–90.
38. M. Iwaki, A. Nakao, M. Kaibara, H. Sasabe, S. Kaneko, H. Nakajima, Y. Suzuki, M. Kusakabe, T. Fujihana, Ion bombardment into inner wall surfaces of tubes and their biomedical applications, *Nucl. Inst. Meth. Phys. Res.* B106 (1995) 618–623.
39. S. Klaumunzer, Q.Q. Zhu, W. Schnabel, G. Schumacher, Ion-beam-induced crosslinking of polystyrene – still an

unsolved puzzle, *Nucl. Inst. Meth. Phys. Res.* B116 (1996) 154–158.

40. V.N. Popok, V.B. Odzhaev, I.P. Kozlov, I.I. Azarko, I.A. Karpovich, D.V. Sviridov, Ion beam effects in polymer films: Structure evolution of the implanted layer, *Nucl. Inst. Meth. Phys. Res.* B129 (1997) 60–64.
41. B. Pignataro, M.E. Fragala, O. Puglisi, AFM and XPS study of ion bombarded poly(methyl methacrylate), *Nucl. Inst. Meth. Phys. Res.* B131 (1997) 141–148.
42. M. Georhegan, F. Abel, High resolution elastic recoil detection analysis of polystyrene depth profiles using carbon ions, *Nucl. Inst. Meth. Phys. Res.* B143 (1998) 371–380.
43. R.I. Khaibullin, Y.N. Osin, A.L. Stepanov, I.B. Khaibullin, Synthesis of metal/polymer composite films by implantation of Fe and Ag ions in viscous and solid state silicone substrates, *Nucl. Inst. Meth. Phys. Res.* B148 (1999) 1023–1028.
44. T. Steckenreiter, E. Balanzat, H. Fuess, C. Trautmann, Pyrolytic effects induced by energetic ions in polymers, *Nucl. Inst. Meth. Phys. Res.* B151 (1999) 161–168.
45. V. Svorcik, V. Rybka, J. Vacik, V. Hnatowicz, R. Ochsner, H. Ryssel, Li+ grafting of ion irradiated polyethylene, *Nucl. Inst. Meth. Phys. Res.* B149 (1999) 331–335.
46. A. Toth, T. Bell, I. Bertoti, M. Mohai, B. Zelei, Surface modification of polyethylene by low keV ion beams, *Nucl. Inst. Meth. Phys. Res.* B148 (1999) 1131–1135.
47. A.L. Evelyn, D. Ila, R.L. Zimmerman, K. Bhat, D.B. Poker, D.K. Hensley, C. Klatt, S. Kalbitzer, N. Just, C. Drevet, Ion beam modification of PES, PS and PVC polymers, *Nucl. Inst. Meth. Phys. Res.* B148 (1999) 1141–1145.
48. D.J. Li, F.Z. Cui, H.Q. Gu, Cell adhesion of F+ ion implantation of intraocular lens, *Nucl. Inst. Meth. Phys. Res.* B152 (1999) 80–88.
49. S. Netcheva, P. Bertrand, Surface topography development of thin polystyrene films under low energy ion irradiation, *Nucl. Inst. Meth. Phys. Res.* B151 (1999) 129–134.
50. R.I. Khaibullin, V.A. Zhikharev, Yu.N. Osin, E.P. Zheglov, I.B. Khaibullin, B.Z. Rameev, B. Aktas, Structural and magnetic properties of iron and cobalt implanted silicone polymers, *Nucl. Inst. Meth. Phys. Res.* B166 (2000) 897–902.
51. Y. Zhang, A.C.H. Huan, K.L. Tan, E.T. Kang, Surface modification of poly(tetrafluoroethylene) films by low energy

Ar+ ion-beam activation and UV-induced graft copolymerization, *Nucl. Inst. Meth. Phys. Res.* B168 (2000) 29–39.
52. G. Compagnini, G.G.N. Angilella, A. Raudino, O. Puglisi, Memory effect on ion beam-induced depolymerization of PMMA, *Nucl. Inst. Meth. Phys. Res.* B175–177 (2001) 559–563.
53. M. Ueda, I.H. Tan, R.S. Dallaqua, J.O. Rossi, J.J. Barroso, M.H. Tabacniks, Aluminum plasma immersion ion implantation in polymers, *Nucl. Inst. Meth. Phys. Res.* B206 (2003) 760–766.
54. J. Davenas, P. Thevenard, Models of the hydrogen release from polymers under ion beam irradiation, *Nucl. Inst. Meth. Phys. Res.* B208 (2003) 170–175.
55. N. Singh, A. Sharma, D.K. Avasthi, Effects of high energy (MeV) ion beam irradiation on polyethylene terephtalate, *Nucl. Inst. Meth. Phys. Res.* B206 (2003) 1120–1123.
56. C. Satriano, C. Scifo, G. Marletta, Study of albumin adsorption on ion beam irradiated polymer surfaces, *Nucl. Inst. Meth. Phys. Res.* B166–167 (2000) 782–787.
57. C. Satriano, S. Carnazza, S. Guglielmino, G. Marletta, Surface free energy and cell attachment onto ion-beam irradiated polymer surfaces, *Nucl. Inst. Meth. Phys. Res.* B208 (2003) 287–293.
58. Y. Yokoyama, T. Tsukamoto, T. Kobayashi, M. Iwaki, Immobilization of collagen by ion bombardment, *Nucl. Inst. Meth. Phys. Res.* B206 (2003) 512–516.
59. J.J. Murphy, M. Patel, A.R. Skinner, I.M. Horn, S.J. Powell, P.F. Smith, Volatile evolution from polymer materials induced by irradiation with He++ ions and comparative pyrolysis experiments, *Nucl. Inst. Meth. Phys. Res.* B215 (2005) 423–435.
60. K. Rockova-Hlavackova, V. Svorcik, L. Bacakova, B. Dvorankova, J. Heitz, V. Hnatowicz, Bio-compatibility of ion beam-modified and RGD-grafted polyethylene, *Nucl. Inst. Meth. Phys. Res.* B225 (2004) 275–282.
61. R.K.Y. Fu, I.T.L. Cheung, Y.F. Mei, C.H. Shek, G.G. Siu, P.K. Chu, W.M. Yang, Y.X. Leng, Y.X. Huang, X.B. Tian, S.Q. Yang, Surface modification of polymeric materials by plasma immersion ion implantation, *Nucl. Inst. Meth. Phys. Res.* B237 (2005) 417–421.

62. V. Zaporojtchenko, J. Zekonyte, S. Wille, U. Schuermann, F. Faupel, Tailoring of the PS surface with low energy ions: relevance to growth and adhesion of noble metals, *Nucl. Inst. Meth. Phys. Res.* B236 (2005) 95–102.
63. J. Zekonyte, V. Zaporojtchenko, F. Faupel, Investigation of the drastic change in the sputter rate of polymers at low ion fluence, *Nucl. Inst. Meth. Phys. Res.* B236 (2005) 241–248.
64. M.A. Parada, N. Delalez, A. de Almeida, C. Muntele, I. Muntele, D. Ila, Low energy ion beam induced changes in ETFE polymer, *Nucl. Inst. Meth. Phys. Res.* B242 (2006) 550–552.
65. A. Turos, A.M. Abdul-Kader, D. Grambole, J. Jagielski, A. Piatkowska, N.K. Madi, M. Al-Maadeed, The effects of ion bombardment of ultra-high molecular weight polyethylene, *Nucl. Inst. Meth. Phys. Res.* B249 (2006) 660–664.
66. J. Wang, J.Y. Chen, P. Yang, Y.X. Leng, G.J. Wan, H. Sun, A.S. Zhao, N. Huang, P.K. Chu, In vitro platelet adhesion and activation of polyethylene terephthalate modified by acetylene plasma immersion ion implantation and deposition, *Nucl. Inst. Meth. Phys. Res.* B242 (2006) 12–14.
67. T. Yotoriyama, A. Nakao, Y. Suzuki, T. Tsukamoto, M. Iwaki, Analysis of cell-adhesion surface induced by ion-beam irradiation into biodegradable polymer, *Nucl. Inst. Meth. Phys. Res.* B242 (2006) 51–54.
68. K.H. Wong, M. Zinke-Allmang, W.K. Wan, J.Z. Zhang, P. Hu, Low energy oxygen ion beam modification of the surface morphology and chemical structure of polyurethane fibers, *Nucl. Inst. Meth. Phys. Res.* B243 (2006) 63–74.
69. C. Ranganathaiaha, G. Shariff, D.K. Avasthi, Carbon-ion-induced modifications of the diffusion kinetics in poly (ethylene terephthalate): a free volume study, *Radiat. Measure.* 36 (2003) 629–634.
70. L. Calcagno, G. Compagnini, G. Foti, Ion-beam effects on optical and reological properties of polystyrene, *Phys. Rev.* B 46(17) (1992) 10573–10578.
71. W. Schnabel, S. Klaumunzer, The effects of ion-beam irradiation of polymers, *Radiat. Phys. Chem.* 37 (1991) 131–134.
72. R. Kumar, R. Prasad, 70 MeV Carbon C^{5+} ion induced modification in polystyrene by positron annihilation, *Radiat. Measure.* 40 (2005) 750–753.

73. K. Proskova, V. Svorcik, V. Rybka, V. Hnatowicz, Selected degradation reactions in polyethylene irradiated with Ar+ and Xe+ ions, *Radiat. Phys. Chem.* 58 (2000) 153–156.
74. V. Svorcik, V. Hnatowicz, P. Stopka, L. Bacakova, J. Heitze, R. Ochsner, H. Ryssel, Amino acids grafting of Ar+ ions modified PE, *Radiat. Phys. Chem.* 60 (2001) 89–93.
75. J.J. Murphy, M. Patel, S.J. Powell, P.F. Smith, Volatile evolution induced by energetic He++ ions in a poly(ester) based polyurethane, *Radiat. Phys. Chem.* 63 (2002) 101–108.
76. M. Veres, M. Fule, S. Toth, I. Pocsik, M. Koos, A. Toth, M. Mohai, I. Bertoti, Raman scattering of ultra-high molecular weight polyethylene treated by plasma-based ion implantation, *Thin Solid Films* 482 (2005) 211–215.
77. H. Tsuji, H. Satoh, S. Ikeda, N. Ikemoto, Y. Gotoh, J. Ishikawa, Surface modification by silver-negative-ion implantation for controlling cell-adhesion properties of polystyrene, *Surf. Coat. Technol.* 103–104 (1998) 124–128.
78. K. Sahre, K.-J. Eichhorn, F. Simon, D. Pleul, A. Janke, G. Gerlach, Characterization of ion-beam modified polyimide layers, *Surf. Coat. Technol.* 139 (2001) 257–264.
79. M. Guenther, G. Gerlach, G. Suchaneck, K. Sahre, K.-J. Eichhorn, B. Wolf, A. Deineka, L. Jastrabik, Ion-beam induced chemical and structural modification in polymers, *Surf. Coat. Technol.* 158–159 (2002) 108–113.
80. T.L. Schiller, D. Sheeja, D.R. McKenzie, D.G. McCulloch, D.S.P. Lau, S. Burn, B.K. Tay, Plasma immersion ion implantation of poly(tetrafluoroethylene), *Surf. Coat. Technol.* 177–178 (2004) 483–488.
81. D.R. McKenzie, K. Newton-McGee, P. Ruch, M.M. Bilek, B.K. Gan, Modification of polymers by plasma-based ion implantation for biomedical applications, *Surf. Coat. Technol.* 186 (2004) 239–244.
82. K.G. Kostov, M. Ueda, I.H. Tan, N.F. Leite, A.F. Beloto, G.F. Gomes, Structural effect of nitrogen plasma-based ion implantation on ultra-high molecular weight polyethylene, *Surf. Coat. Technol.* 186 (2004) 287–290.
83. M. Ueda, K.G. Kostov, A.F. Beloto, N.F. Leite, K.G. Grigorov, Surface modification of polyethylene terephthalate by plasma immersion ion implantation, *Surf. Coat. Technol.* 186 (2004) 295–298.

84. J. Wang, N. Huang, C.J. Pan, S.C.H. Kwok, P. Yang, Y.X. Leng, J.Y. Chen, H. Sun, G.J. Wan, Z.Y. Liu, P.K. Chu, Bacterial repellence from polyethylene terephthalate surface modified by acetylene plasma immersion ion implantation deposition, *Surf. Coat. Technol.* 186 (2004) 299–304.
85. J. Zhang, X. Zhang, H. Zhou, Surface-restructuring behavior of aged PTFE irradiated by a high-flux nitrogen ion beam, *Surf. Coat. Technol.* 187 (2004) 250–256.
86. A.M. Ektessabi, K. Yamaguchi, Changes in chemical states of PET films due to low and high energy oxygen ion beam, *Thin Solid Films* 377–378 (2000) 793–797.
87. M. Veres, M. Fule, S. Toth, I. Pocsik, M. Koos, A. Toth, M. Mohai, I. Bertoti, Raman scattering of ultra-high molecular weight polyethylene treated by plasma-based ion implantation, *Thin Solid Films* 482 (2005) 211–215.
88. A. Turos, J. Jagielski, A. Piatkowska, D. Bielinski, L. Slusarski, N.K. Madi, Ion beam modification of surface properties of polyethylene, *Vacuum* 70 (2003) 201–206.
89. W. Shi, X.Y. Li, H. Dong, Improved wear resistance of ultra-high molecular weight polyethylene by plasma immersion ion implantation, *Wear* 250 (2001) 544–552.
90. N. Gavrilov, D. Yakusheva, A. Kondyurin, Structure of polyethylene after pulse ion beam treatment, *J. Appl. Polym. Sci.* 69 (1998) 1071–1077.
91. A. Kondyurin, V. Romanova, V. Begishev, I. Kondyurina, R. Guenzel, M. Maitz, Crosslinked polyurethane coating on vascular stents for enhanced X-ray contrast, *J. Bioact. Compat. Polym.* 20 (2005) 77–93.
92. A.V. Kondyurin, M.F. Maitz, V.A. Romanova, V.P. Begishev, I.V. Kondyurina, R. Guenzel, Drug release from polyureaurethane coating modified by plasma immersion ion implantation, *J. Biomater. Sci. Polym. Edn.* 15(2) (2004) 145–159.
93. R.I. Khaibullin, B.Z. Rameev, V.N. Popok, E.P. Zheglov, A.V. Kondyurin, V.A. Zhikharev, B. Aktas, An influence of the viscosity of polymer substrate on ion beam synthesis of iron granular films, *Nucl. Inst. Meth. Phys. Res.* B206 (2003) 1115–1119.
94. A. Kondyurin, P. Volodin, J. Weber, Plasma immersion ion implantation of Pebax polymer, *Nucl. Inst. Meth. Phys. Res.* B251 (2006) 407–412.

95. A. Kondyurin, B.K. Gan, M.M.M. Bilek, K. Mizuno, D.R. McKenzie, Etching and structural changes of polystyrene films during plasma immersion ion implantation from argon plasma, *Nucl. Inst. Meth. Phys. Res.* B251 (2006) 413–418.
96. B.K. Gan, M.M.M. Bilek, A. Kondyurin, K. Mizuno, D.R. McKenzie, Etching and structural changes in nitrogen plasma immersion ion implanted polystyrene films, *Nucl. Inst. Meth. Phys. Res.* B 247 (2006) 254–260.
97. G.A. Mesyats, Yu.S. Klyachkin, N.V. Garilov, V.N. Mizgulin, R.M. Yakushev, A.V. Kondyurin, Ion beam modification of polyethylene and adhesion to epoxy adhesive, *Vacuum* 47(9) (1996) 1085–1087.
98. G. Mesyats, Yu. Klyachkin, N. Gavrilov, A. Kondyurin, Adhesion of Polytetrafluorethylene modified by an ion beam, *Vacuum* 52 (1999) 285–289.
99. A. Kondyurin, V. Karmanov, R. Guenzel, Plasma immersion ion implantation of polyethylene, *Vacuum* 64 (2002) 105–111.
100. A. Kondyurin, R. Khaybullin, N. Gavrilov, V. Popok, Pulse and continuous ion beam treatment of polyethylene, *Vacuum* 68 (2003) 341–347.
101. N. Koshida, Y. Suzuki, T. Aoyama, Low energy ion implantation studies of polyacetylene films, *Nucl. Instr. Meth.* B37–38 (1989) 708–711.
102. O.Yu. Posudievski, N.G. Myasnikova, A.A. Chuiko, Growth of electroconductivity of polymers, caused by implantation of molecular ions, *Dokladi Akademii Nauk*, 324 (1992) 131–132.
103. A.N. Aleshin, A.V. Gribanov, A.V. Dobrodumov, A.V. Suvorov, I.S. Shlimak, Electrophysical properties of polyimide PM films treated by ion bombardment, *Solid State Physics* 31(1) (1989) 12–18.
104. A.N. Aleshin, A.V. Suvorov, Influence of electron-electron interactions on low temperature conductivity of polyimide films irradiated by ions, *Solid State Physics*, 32(8) (1990) 1717–1720.
105. L. Guzman, R. Celva, A. Miotello, E. Voltolini, F. Ferrari, M. Adami, Polymer surface modification by ion implantation and reactive deposition of transparent films, *Surf. Coat. Technol.*, 103–104 (1998) 375–379.

106. V. Svorcik, V. Rybka, K. Volka, V. Hnatowicz, J. Kvitek, P. Seidl, Ion implantation into polypropylene, *Jpn. J. Appl. Phys.* 31 (1992) L287–390.
107. T. Sasuga, S. Kawanishi, M. Nishii, T. Seguchi, I. Kohno, Effects of ion irradiation on the mechanical properties of several polymers, *Radiat. Phys. Chem.* 37(1) (1991) 135–140.
108. M. Guenther, G. Gerlach, G. Suchaneck, K. Sahre, K.-J. Eichhorn, V. Baturin, S. Duvanov, Physical properties and structure of thin ion-beam modified polymer film, *Nucl. Inst. Meth. Phys. Res. Section B: Beam Interactions with Materials and Atoms*, 216 (2004) 143–148.
109. T.L. Schiller, D. Sheeja, D.R. McKenzie, D.G. McCulloch, D.S.P. Lau, S. Burn, B.K. Tay, Plasma immersion ion implantation of poly(tetrafluoroethylene), *Surf. Coat. Technol.* 177–178 (2004) 483–488.
110. A. Valenza, A.M. Visco, L. Torrisi, N. Campo, Characterization of ultra-high-molecular-weight polyethylene (UHMWPE) modified by ion implantation, *Polymer* 45(5) (2004) 1707–1715.
111. F.F. Komarov, A.V. Leontyev, V.V. Grigoryev, M.A. Kamishan, Ion implantation for local change of the optical constants of polymer films, *Nucl. Inst. Meth. Phys. Res. Section B: Beam Interactions with Materials and Atoms* 191(1–4) (2002) 728–732.
112. K. Krezhov, K. Velitchkova, S. Balabanov, Transport properties of selenium implanted polymer composites, *Vacuum* 69(1–3) (2002) 113–118.
113. J.M. Colwell, E. Wentrup-Byrne, J.M. Bell, L.S. Wielunski, A study of the chemical and physical effects of ion implantation of micro-porous and nonporous PTFE, *Surf. Coat. Technol.* 168(2–3) (2003) 216–222.
114. R.J. Rodríguez, J.A. García, R. Sánchez, A. Pérez, Blas Garrido, J. Morante, Modification of surface mechanical properties of polycarbonate by ion implantation, *Surf. Coat. Technol.* 158–159 (2002) 636–642.
115. H. Tsuji, M. Izukawa, R. Ikeguchi, R. Kakinoki, H. Sato, Y. Gotoh, J. Ishikawa, Improvement of polydimethylsiloxane guide tube for nerve regeneration treatment by carbon negative-ion implantation, *Nucl. Inst. Meth. Phys. Res.*

Section B: Beam Interactions with Materials and Atoms 206 (2003) 507–511.

116. Y. Suzuki, M. Iwaki, S. Tani, G. Oohashi, M. Kamio, Ion implantation into ePTFE for application of a dural substitute, *Nucl. Inst. Meth. Phys. Res. Section B: Beam Interactions with Materials and Atoms* 206 (2003) 538–542.
117. H. Tsuji, H. Sasaki, H. Sato, Y. Gotoh, J. Ishikawa, Neuron attachment properties of carbon negative-ion implanted bioabsorbable polymer of poly-lactic acid, *Nucl. Inst. Meth. Phys. Res. Section B: Beam Interactions with Materials and Atoms* 191(1–4) (2002) 815–819.
118. K. Dworecki, T. Hasegawa, K. Sudlitz, S. Wsik, Modification of electrical properties of polymer membranes by ion implantation, *Nucl. Inst. Meth. Phys. Res. Section B: Beam Interactions with Materials and Atoms* 166–167 (2000) 646–649.
119. B. Ranby, J.F. Rabek, *Photodegradation, Photo-oxidation and Photostabilization of Polymers*, New York, Wiley, 1975.
120. N.M. Emanuel, A.L. Buchachenko, *Chemical Physics of Polymer Degradation and Stabilization*, Brill Academic Publishers, 1987, p. 336.

4 Structure of polymers after ion beam treatment

The treatment of polymers by ion beam changes their color from a fluence of 10^{14} to 10^{15} ions/cm^2. At such a low fluence of ion beam implantation, the surfaces of polyethylene, PET, polystyrene and other polymers become rainbow-like. At fluences higher than 10^{15} ions/cm^2 the surface becomes dark yellow and changes to brown with fluence increase (Fig. 4.1). At a fluence of 10^{16} ions/cm^2 the surface becomes dark or silver-like. Such color changes can be used for initial visual analysis of the surface treatment effect and for visual analysis of fluence distribution on the sample surface. The color changes are observed for all kinds of polymers, but mainly strong changes are observed for polymers with high carbonization ability such as polyethylene and polystyrene. For depolymerizing polymers under irradiation, the color becomes lighter. For example, PTFE changes color from white to gray under ion beam only with high fluence. PLGA film becomes a milk-like color in reflection and light brown in transmittance at high fluence of treatment. But at low fluence it is not possible to gauge whether PTFE and PLGA have been treated or not.

The color change is observed only in the thin surface layer. The back side of polymer films does not change. The dark area corresponds to penetration of the ions into the polymer. For example, a modified layer of PET after PIII with nitrogen ions of 20 keV energy is viewed at a distance of about 100 nm from the surface (Fig. 4.2). The boundary between the modified and bulk unmodified layer

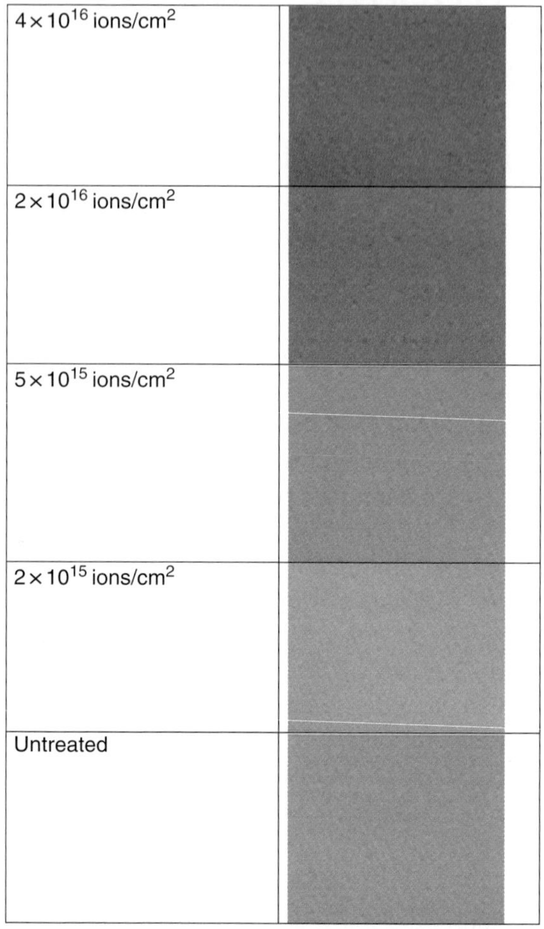

Fig. 4.1. Optical image of polyethylene samples treated with different fluence of PIII.

is not sharp. The dark color decreases smoothly to white in the unmodified layer. Variation of the modified layer thickness is significant in comparison with the thickness of the layer itself. The color of the modified surface is stable and cannot be changed by solvents without completely removing the whole modified layer.

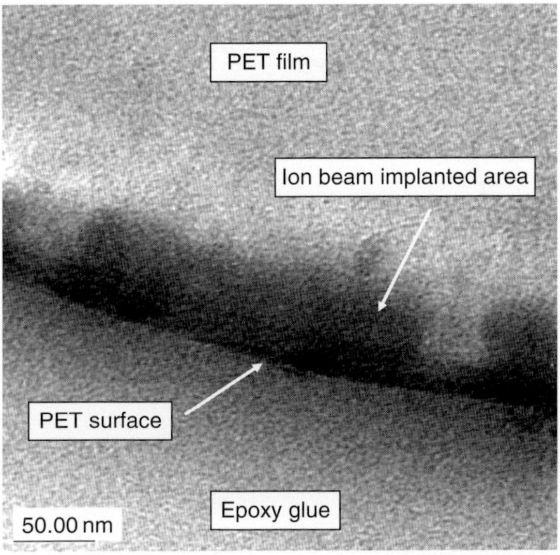

Fig. 4.2. Transmission electron microscope (TEM) image of cross-section of PET after PIII by nitrogen ions with 20 keV energy (University of Sydney, Australia) in epoxy matrix. The TEM photo was taken by B.K. Gan.

The changes in the modified layer are observed in transmission UV visual spectra. The UV spectra of untreated polymer film shows peaks of light absorbance corresponding to energetic levels of valence electrons in polymer macromolecules. For most kinds of polymers, the spectra do not contain any absorbance lines in the visual region of the light. The absorbance in the UV region depends on the presence of chromophoric groups like carbonyl or aromatic ring. In the spectra of ion beam implanted polymer the additional absorbance appears as a shoulder which is stronger in short wavenumber and weaker in long wavenumber (Fig. 4.3). The absorbance is caused its appearance in carbonized structures: unsaturated carbon=carbon double bonds in macromolecules, aromatic ring structures and its condensed structures up to graphite-like planes. The absorbance increases with fluence and with energy of ions: more energy is transferred to the surface polymer layer.

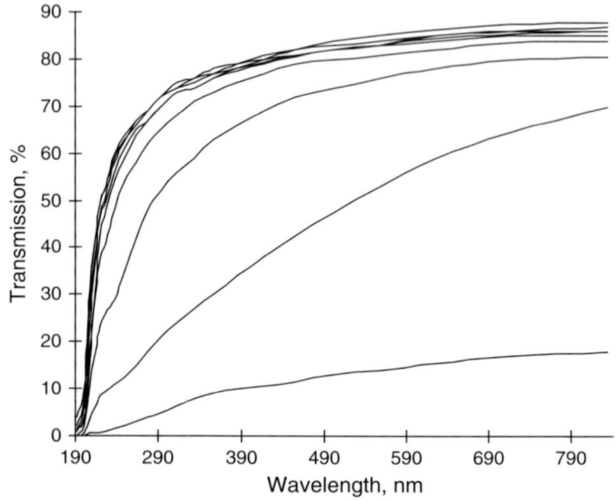

Fig. 4.3. UV-vis transmission spectra of polyethylene (LDPE) after PIII (Forschungszentrum Rossendorf, Dresden, Germany) with fluence (from top to bottom): 0, 10^{13}, 5×10^{13}, 10^{14}, 5×10^{14}, 10^{15}, 5×10^{15}, 10^{16}, 10^{17} ions/cm^2.

This results in a higher number of carbonized structures (absorbance becomes stronger) and longer connected unsaturated structures (absorbance shifts to the long wavelength region). The absorbance does not depend on the post-atmosphere environment and it is stable with time after treatment. The discolorization of the polymer after PIII and ion beam implantation by oxygen, carbon, nitrogen and argon ions was not observed as it was observed for H+ irradiated polyimide and Plexiglas stored in light and air after implantation [1]. Therefore, the optical absorbance of polymers after ion beam implantation can be used for the calibration of treatment fluence by UV-vis spectra.

For example, Fig. 4.4 shows the absorbance dependence of polyethylene film at separate wavelengths of the light. With fluence increase the absorbance curve rises. The curve is non-linear in both the linear and logarithm scales. Despite its complex shape, the curve can be used for fluence calibration for unknown equipment or a new experimental

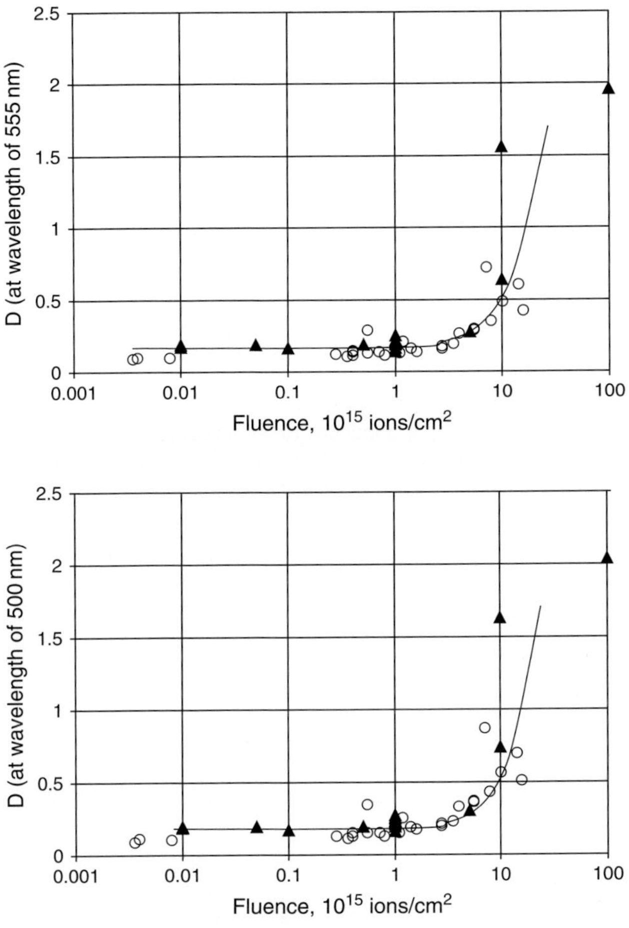

Fig. 4.4. UV-vis optical density (absorbance) of polyethylene film after ion beam implantation with measured current density and treatment fluence (full signs). The implantation was done on a Pulsar ion beam implanter (Institute of Technical Chemistry, Perm, Russia). Empty signs correspond to PIII treated polyethylene with calculated fluence of treatment (Forschungszentrum Rossendorf, Dresden, Germany and University of Sydney, Australia).

setup for PIII or ion beam implantation. This curve was done for polyethylene film samples which were treated in the ion beam implanter with measured ion current density. The ion current density was measured using a Faraday cap.

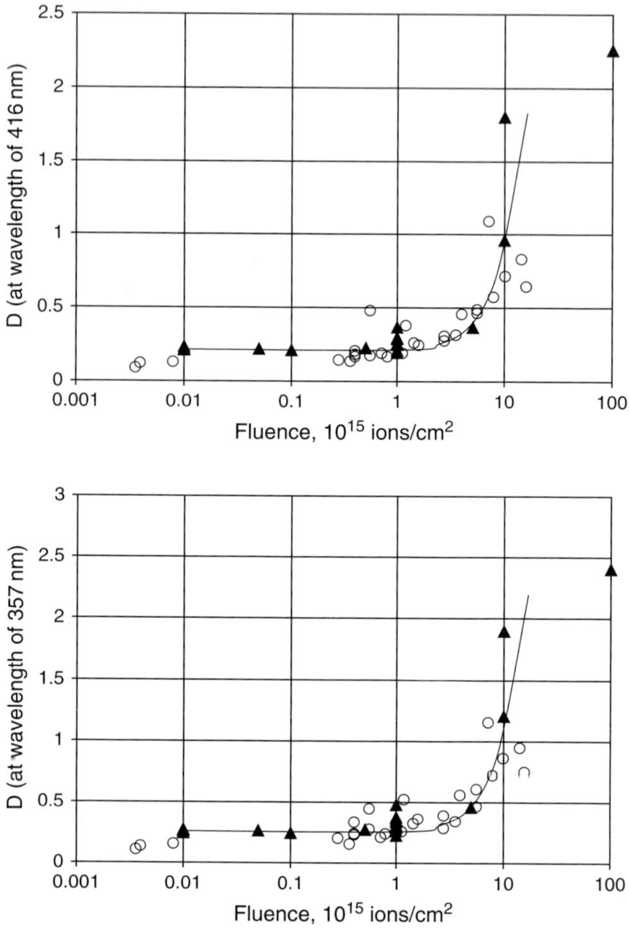

Fig. 4.4. (Continued)

This curve is reliable for low density polyethylene (LDPE) film of 50 μm thickness treated by nitrogen ions with an energy of 20 keV in a pulse periodical regime with 300 μs pulse duration and 1 Hz frequency of pulse repetition. The calibration curve does not have any correction for electrical charge of the polyethylene film during ion beam implantation, because the film is thin enough and electrical breakdown processes could neglect the charge effect.

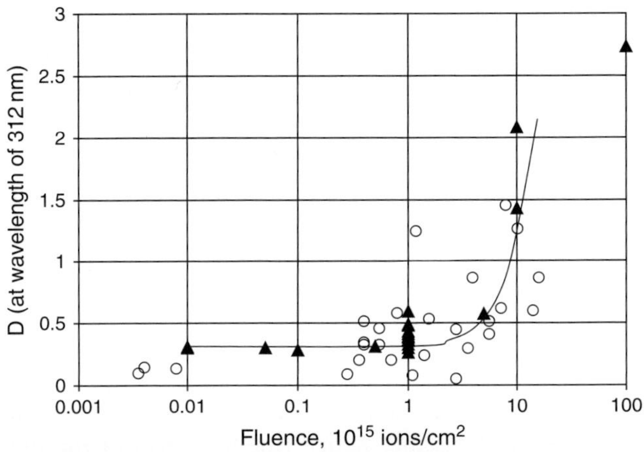

Fig. 4.4. (Continued)

It is essential that the ion beam implantation process must not deform or melt the polyethylene which means that a non-thermal ion beam implantation regime must be used. The calibration curves for absorbance at 357, 416, 500 and 555 nm give different calibration coefficients which may be used to estimate the deviation of the fluence calculations. Different wavelengths of UV spectra have different sensitivities to fluence dependence. The region of short wavelength (250–350 nm) is more sensitive to low fluence. The long wavelength region (500–600 nm) is preferable for the estimation of high fluence of ion beam implantation.

If the energy of ions is changed the curve would not show right fluence. The fluence that is dependent on current density has to be recalculated. In some cases, for example for plasma immersion ion implantation or similar ion beam sources, the current density of ion beam corresponds roughly to the law:

$$I \approx U^{3/2} \tag{4.1}$$

where I is the current and U is the bias applied to high voltage electrodes. The increase of ion energy, the ion current and the corresponding fluence of ion beam implantation are increased by this law. For a more accurate calculation, the current density distribution based on plasma sheet dynamics must be recalculated for a different bias voltage.

If another kind of ion is used, the plasma density could be used for fluence recalculation. In such a case, the plasma density must be measured for each kind of ion, including ions used for calibration. For a rough estimation, the fluence could be recalculated proportionally to the plasma density:

$$\frac{F_1}{d_1} = \frac{F_2}{d_2} \tag{4.2}$$

where F_1 and d_1 are fluence and plasma density of the 1st kind of ions and F_2 and d_2 are fluence and plasma density of the 2nd kind of ions.

The same recalculation with plasma density could be done, if the pressure of the working gas and configuration of plasma electrodes or magnetic field are changed, if they are presented. After measuring the plasma density, the new fluence of ion beam implantation could be recalculated using equation (4.2).

The mostly difficult variant for fluence estimation is when the configuration of the high voltage electrode is changed. In such a case, the calibration curve must be repeated, because the calculation of plasma sheet dynamics must be done for fluence estimation, which is much more difficult than recalculating the calibration curve.

If the pulse parameters of bias such as pulse duration or pulse frequency are changed, the fluence of ion beam implantation could be recalculated proportionally to the pulse duration and pulse frequency:

$$\frac{F_1}{\tau_1 f_1} = \frac{F_2}{\tau_2 f_2} \tag{4.3}$$

where F_i is the fluence, f_i is the frequency and τ_i is the pulse duration of the i-regime of ion beam implantation.

Such a calibration procedure can give an estimation of fluence treatment with an accuracy of $\pm 5\%$. By our experience, a more accurate measurement is nonsensical, due to the errors associated with variations in the plasma density, the plasma sheet (due to charge accumulation), the quality and composition of polymer surface layer, and inaccuracies in the structure measurement.

The absorbance of UV-visible light of modified polymer depends on the type of polymer, and specifically, its degree of carbonisation in response to ion beam implantation. For example, polyvinyl chloride has less carbonization ability than polyethylene and less absorbance at the same fluence of ion beam implantation (Fig. 4.5). The difference appears due to the presence of a chlorine atom in the polyvinyl chloride (PVC) macromolecule. In the ion beam implantation process, the implanted ions can transfer energy to the chlorine atoms which accept more energy due to bigger mass, than carbon, and spattered chlorine atoms can easy release as Cl_2 and HCl gases from treated polymer. Therefore, the areas with implanted ions loose energy which may

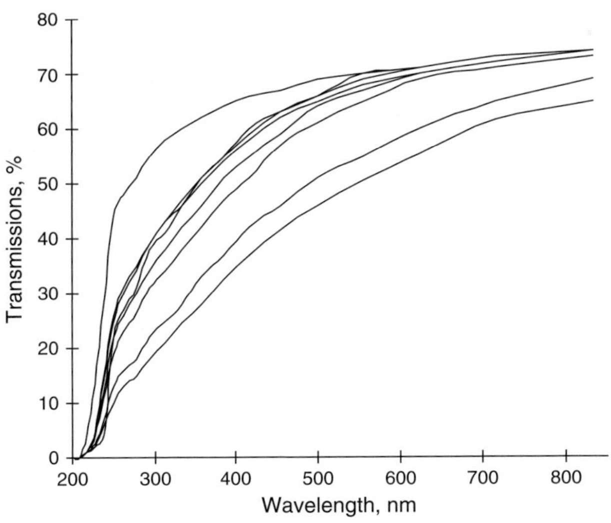

Fig. 4.5. UV-vis transmission spectra of polyvinylchloride (PVC) after PIII (Forschungszentrum Rossendorf, Dresden, Germany) with fluence (from top to bottom): 0, 10^{13}, 5×10^{13}, 10^{14}, 5×10^{14}, 10^{15}, 5×10^{15}, 10^{16} ions/cm^2.

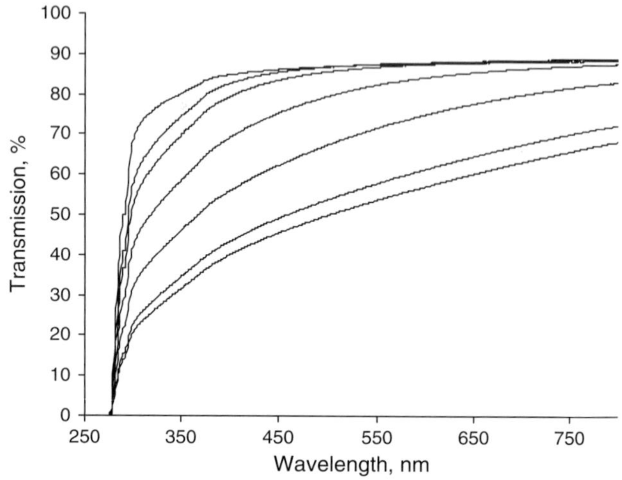

Fig. 4.6. UV-vis transmission spectra of polystyrene sheet after PIII by 20 keV nitrogen ions with fluence (from top to bottom): 0, 5×10^{14}, 10^{15}, 2×10^{15}, 5×10^{15}, 2×10^{16}, 4×10^{16} ions/cm^2. The treatment was done at the University of Sydney, Australia.

have potentially been use to spatter the carbon atom, and hence the effect of carbonization is reduced. Polystyrene (PS) has more absorbance than PVC, because it does not have chlorine atoms in the macromolecule (Fig. 4.6). However, polystyrene has less absorbance than polyethylene due to the presence of an aromatic ring. The aromatic ring is placed on the side of the macromolecule backbone. At collision and breaking of the C—C bond between the backbone macromolecule and the aromatic ring, the radical of the aromatic ring becomes free and it can evaporate from the polymer into the vacuum. The polystyrene surface layer accepts ion energy from the ion beam but loses part of the macromolecule without carbonization. Polyethylene terephthalate (PET) (Fig. 4.7) has oxygen atoms in the polymer backbone, but carbonization is less than for polyethylene and polyvinyl chloride, and the absorbance is week even at high fluence of ion beam implantation. Thus, with an increasing in the available gaseous products at destruction

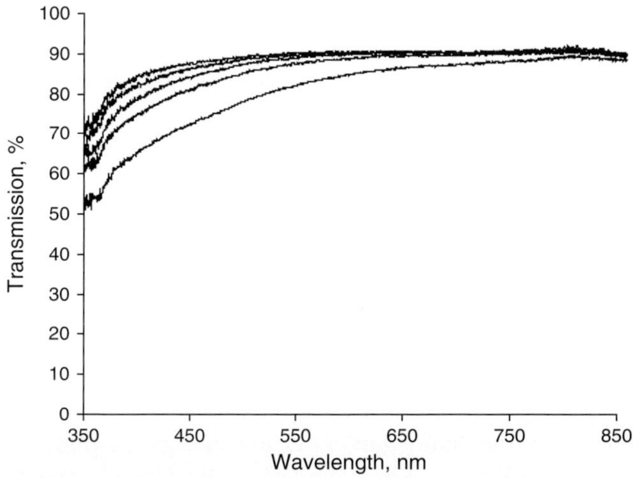

Fig. 4.7. UV-vis transmission spectra of polyethylene terephthalate (PET) after PIII by 20 keV nitrogen ions (University of Sydney, Australia) with fluence (from top to bottom): 0, 5×10^{14}, 10^{15}, 2×10^{15}, 5×10^{15} ions/cm^2.

of the macromolecule in ion beam, the carbonization effect decreases. Examples of the lowest carbonized polymers are polylacticglycolic acid (PLGA) and polytetrafluorethylene (PTFE), which are mostly decomposed under any kind of irradiation. However, such polymers could be carbonized under ion beam with some amount of carbonized fraction.

The appearance of unsaturated carbon—carbon bonds in ion beam implanted polymer in the form of diene structures and condensed aromatic structures occurs in the local field by track of ion penetration. The localization of structure defects with high density leads to an appearance of unsaturated carbon clusters. The number and size of these clusters depend on the fluence and energy of the implanting ions, as well as the polymer type. Under high fluence, there is a high concentration of structural defects resulting in a larger number of carbon atoms in each cluster and consequently a bigger cluster size. However, even at the highest fluences the continuous carbon clusters are not observed in any irradiated polymers. The carbon clusters have island structures with separation by sp^3 bond structures, oxygen-containing groups and residual hydrogen-containing groups. The presence of carbon clusters significantly decreases the energy gap between the conduction and valence bands in irradiated polymers. Consequence, less energy is required for electrons to make the transition to the conduction band and the conductivity of the irradiated polymer is far higher. The bandgap width may be calculated using Tauc's formula for semiconductors:

$$D(E) = B\frac{(E - E_g)^2}{E} \tag{4.4}$$

where E_g is the band gap, E is the energy of photon and B is the coefficient. For calculation the UV-vis spectra is presented in Tauc coordinates:

$$y = \sqrt{D \cdot k} \text{ versus } k \tag{4.5}$$

where $k(\mathrm{cm}^{-1})$ is the wavenumber and D is the optical density (absorbance) calculated from UV-vis spectra. The linearization of the curve (4.4) gives coefficients for the calculation of E_g (Fig. 4.8). We have to note that the electronic band gap value calculated from UV spectra by equation (4.4) could be used only for estimation of the real electronic structure of irradiated polymer. This method does not consider distribution of the clusters by size and regularity of the clusters as well as distribution on depth in the polymer surface layer, which have an influence on energetic levels of the π-electron system of the irradiated polymer. Despite of absence of structure detalization, the change of average band gap for the surface layer of the polymer irradiated under different regimes of ion beam implantation could be estimated.

The usual band gap dependence on fluence of implantation has an asymptotical curve which is described by the exponential function:

$$E_g = E_{g\infty} + (E_{g0} - E_{g\infty}) \cdot e^{-\varepsilon D} \tag{4.6}$$

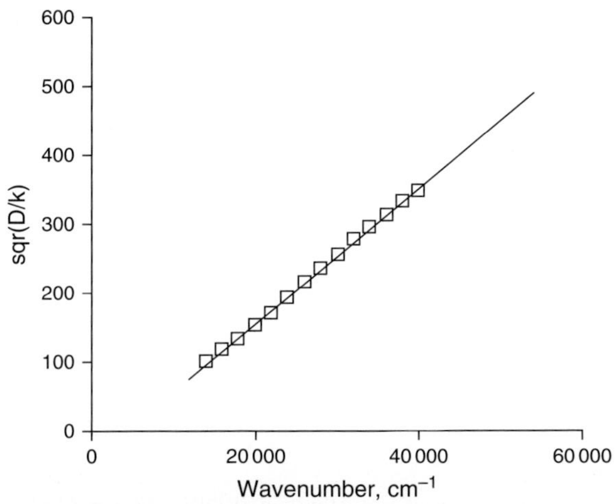

Fig. 4.8. UV-vis spectra of polyethylene (LDPE) after ion beam implantation by 20 keV nitrogen ions (ion beam implanter ILU-4, Physical-technical Institute, Kazan, Russia). Spectrum is presented in Tauc coordinates.

where $E_{g\infty}$ is the band gap of the completely carbonized polymer layer, ε is the specific growth rate of carbon clusters in terms of implantation fluence and E_{g0} is the band gap of untreated polymer. An example of a fluence dependent band gap for polyethylene under nitrogen ion beam is presented in Fig. 4.9. The band gap of initial untreated polymer is about 2–3 eV for different kinds of polymers, which corresponds to isolators with low conductivity ($\rho = 10^{12}$–10^{15} Ohm · cm). The modified polymers with high fluence of implantation have a band gap closed to amorphous carbon or graphite at 0.6 eV. The specific conductivity of the implanted layer can reach up to 1–100 Ohm • cm.

The carbonization of the surface layer is well observed by Raman spectra. The effect of Raman spectra is based on scattering of laser monochromatic light on molecular vibrations. The intensity of Raman lines is proportional to the derivative of quadropole moment of the scattering molecule on normal vibration coordinates. The effect is that

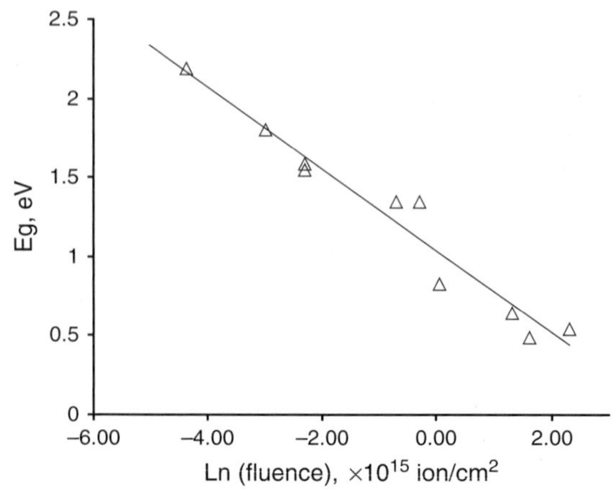

Fig. 4.9. Energetic gap of polyethylene (LDPE, 50 mm film) after ion beam implantation in continuous regime (ion beam implanter ILU-4, Physical-technical Institute, Kazan, Russia). The experiment was done in cooperation with R. Khaibullin.

the intensity is higher for non-polar fragments of molecules than for polar. An advantage of Raman spectra is to get high intensity of vibrational lines in carbonized structures like diene and aromatic ring.

The modified layer of polymer after ion beam implantation is very thin and it is difficult to get Raman spectra from the thin implanted layer on the top of the unmodified polymer. Usually, good quality Raman spectra from the thin layer of polymer can be recorded using an optical microscope, when the exciting laser beam is focused on the thin surface layer of the polymer. The scattered light can be collected in reflection geometry of the laser beam and the laser beam can be focused on a field of some tens of μm. The Raman spectra in the visual region of the exciting laser beam gives the best result in comparison with the exciting laser beam in the near infrared region, due to easy overheating (even burning) of the polymer by high intensity laser beam and high absorbance of the carbonized layer.

In spite of this, if Raman spectra are excited by laser line in the visual light region, the laser beam excites the energetic levels of the π-electron in carbon clusters which is observed in UV-vis spectra of modified polymers by absorbance in electron energetic levels. Therefore, the Raman spectra have resonance excitation and the intensity of Raman spectra increases to 50–230 times in comparison with ordinary Raman spectra. This resonance effect gives significant amplification of the Raman signal of the thin carbonized layer which is stronger than the Raman signal from bulk polymer.

For example, the Raman spectra of polyolefin elastomer EG-8200 (co-polymer of ethylene and octene-1) after PIII are presented in Fig. 4.10. Presented spectra were recorded using a microscope attachment which gives a focus of the excitation laser beam in the thin layer of polymer surface. However, the focus spot of the laser beam is larger than the

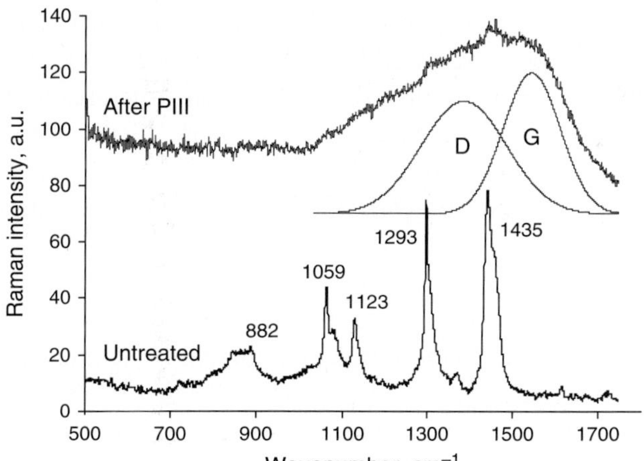

Fig. 4.10. Micro-Raman spectra of ethylene octane-1 co-polymer (EG-8200©) before and after PIII by 20 keV nitrogen ions (Forschungszentrum Rossendorf, Dresden, Germany). Fitting peaks G and D are shown for Raman spectra of implanted polymer. Micro-Raman spectra were recorded in a backscattering mode excited by Nd:YAG laser irradiation (2ω, $\lambda = 532.14$ nm) on a diffraction double monochromator spectrometer HR800, Jobin Yvon with LabRam System 010. Optical microscope was used for focusing the exciting laser beam and for collection of Raman scattered light.

modified layer of polymer and Raman spectra correspond to a mixing of modified and unmodified layers of polymer. With variations of the focus position the spectra of the carbonized layer, transition layer or unmodified (bulk) layer of treated polymer could be recorded. Raman spectra of the initial polymer contain narrow lines of macromolecule vibrations. Polyolefin elastomer EG-8200 has such lines which are observed at 846, 882, 1059, 1123, 1293, 1366 and 1435 cm^{-1}. The spectrum of the sample after PIII does not contain narrow lines. The high intensity broad band from 1000 to 1700 cm^{-1} is observed in Raman spectra. This broad band is attributed to vibrational states in carbon structures. Two peaks could be separated by fitting of the spectra: the G-peak at 1543 cm^{-1} is attributed to vibrations of the E_{2g} mode in graphite structures, where carbon has sp^2 hybridization of valence electrons; and the D-peak at 1384 cm^{-1} is attributed to the A_{1g} mode of graphite vibrations. The last A_{1g} mode becomes visible in Raman spectra of graphite structures which are disturbed by carbon structures with sp^3 hybridization on the edges of the graphite cluster. The ratio of D/G peak intensity and positions of the lines could be used for analysis of carbon structures and calculation of graphitic cluster size.

In comparison with olefin elastomer, the presence of significant amounts of oxygen in the backbone of the polyethylene terephthalate macromolecule increases etching rate and decreases the amount of carbon collected in the surface layer under ion beam. The micro-Raman spectra of PET film shows that vibrational lines of PET macromolecules remain in spectra after high fluence of ion beam implantation (Fig. 4.11). The narrow lines at 633, 858, 1096, 1291, 1615 and 1727 cm^{-1} observed in the spectra of unmodified PET are viewed on the top of wide carbon bands in the spectra of PIII modified PET. The intensity of these lines reduces as the fluence of ion beam implantation increases. However, even at highest fluence of ion beam implantation (10^{17} ions/cm^2) these lines are still observed. In contrast,

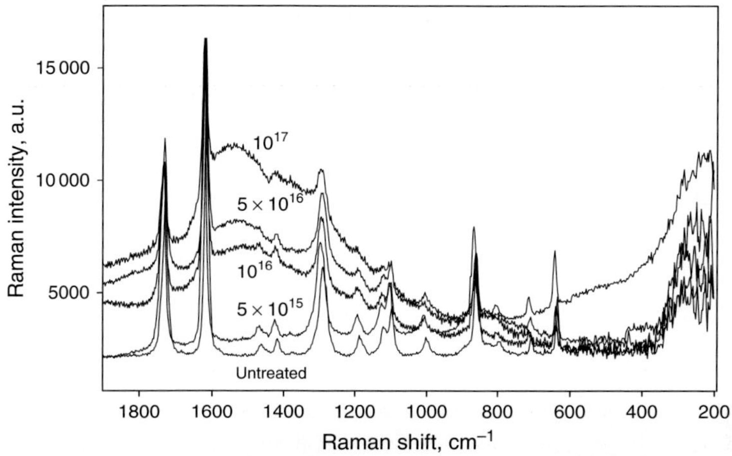

Fig. 4.11. Micro-Raman spectra of polyethylene terephthalate (PET) treated with PIII with 30 keV nitrogen ions (Forschungszentrum Rossendorf, Dresden, Germany). Micro-Raman spectra were recorded in a backscattering mode excited by Nd:YAG laser irradiation (2ω, $\lambda = 532.14$ nm) on a diffraction double monochromator spectrometer HR800, Jobin Yvon with LabRam System 010. Optical microscope was used for focusing the exciting laser beam and for collection of Raman scattered light. Fluence is noted in ions/cm^2.

the intensity of carbon D- and G-peaks increase with fluence of ion beam implantation, but their profiles are unchanged.

Another character of spectra changes is observed at ion beam implantation into polyamide. Micro-Raman spectra of polyamide-6.6 (Ultramid®A) are presented in Fig. 4.12. The spectrum of unmodified polyamide contains narrow lines of macromolecule vibrations. After PIII treatment with low fluence, the spectrum of macromolecular vibrations becomes weaker and a broad band between 1800 and 600 cm^{-1} wavenumbers appears as the main spectral response. With higher fluence of PIII treatment the carbon band becomes more intensive and narrower. The shape of the band is observed as for EG8200 co-polymer and PET up to PIII fluence of 10^{16} ions/cm^2: an intensive peak at 1540 cm^{-1} with low frequency shoulder. At a high fluence of 10^{17} ions/cm^2 the band is split into two narrow lines at

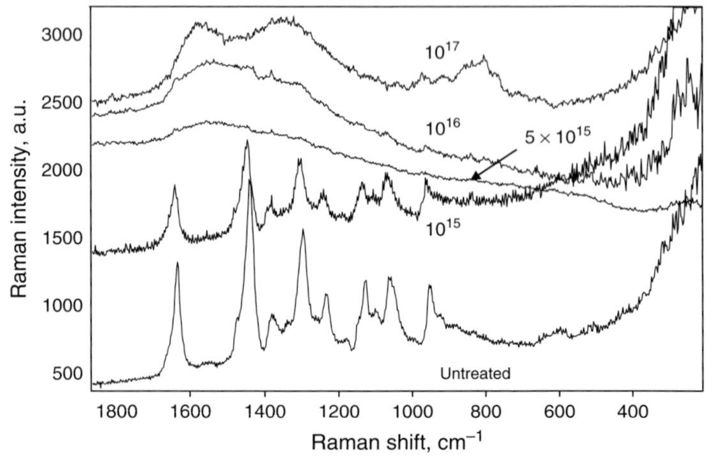

Fig. 4.12. Micro-Raman spectra of polyamide-6.6 (Ultramid) treated with PIII with 30 keV nitrogen ions (Forschungszentrum Rossendorf, Dresden, Germany). Micro-Raman spectra were recorded in a backscattering mode excited by Nd:YAG laser irradiation (2ω, $\lambda = 532.14$ nm) on a diffraction double monochromator spectrometer HR800, Jobin Yvon with LabRam System 010. Optical microscope was used for focusing the exciting laser beam and for collection of Raman scattered light. Fluence is noted in ions/cm^2.

1589 and 1361 cm^{-1}. These lines are attributed to vibrational modes in the carbon structure of the surface modified layer of polyamide.

The doublet of narrow separated lines is observed in Raman spectra of highly carbonized polymers. Such characteristic lines do not appear after low energy implantation (lower than 20 keV) and at low fluence of implantation (lower than 10^{16} ions/cm^2). However, such doublets do not appear in all kinds of polymers. We observed such a doublet only in polymers with a high rate of carbonization, for example polyethylene (Fig. 4.13). The micro-Raman spectra of unmodified polyethylene sometimes contain a high level of fluorescence and vibrational lines could not be observed on the background of wide and intensive fluorescence signal. The spectrum of PIII treated polyethylene

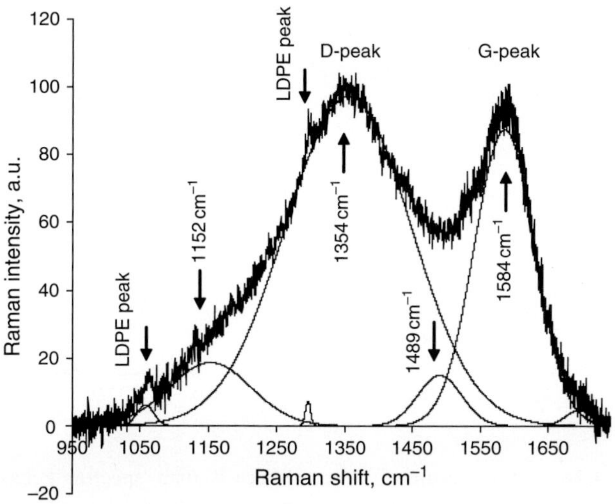

Fig. 4.13. Micro-Raman spectra of polyethylene (LDPE) after PIII with 30 keV energy at 5×10^{16} ions/cm^2 fluence (Forschungszentrum Rossendorf, Dresden, Germany). Micro-Raman spectra were recorded in a backscattering mode excited by Nd:YAG laser irradiation (2ω, $\lambda = 532.14$ nm) on a diffraction double monochromator spectrometer HR800, Jobin Yvon with LabRam System 010. Optical microscope was used for focusing the exciting laser beam and for collection of Raman scattered light.

contains clear observed lines of carbonized structure at 1588 and 1353 cm^{-1}.

The influence of ion energy on carbonized structures is well observed on an example of PIII treated Pebax co-polymer. The micro-Raman spectra of Pebax co-polymer are presented in Figs 4.14 and 4.15. The spectra correspond to Pebax co-polymer treated with the same fluence, kind of ions and pulse duration and frequency of pulse repetition. Only the energy of implanting ions is different. At low energy, the Raman spectrum contains a graphitic doublet and narrow peaks of macromolecule vibrations. At high energy of ions, the graphitic structure dominates the macromolecular vibrational lines and carbon G- and D-peaks become separated.

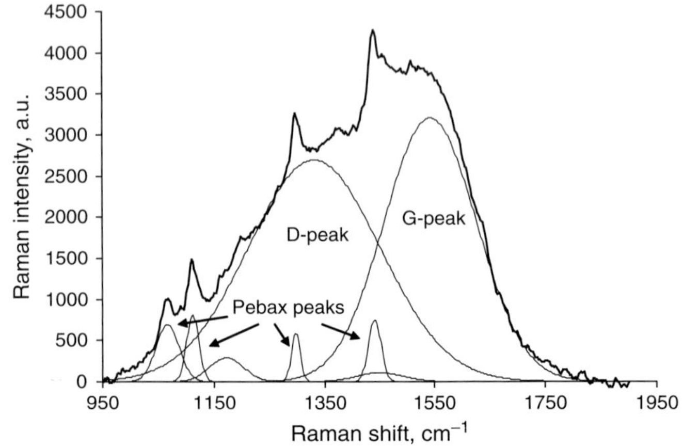

Fig. 4.14. Fitting result of Pebax Micro-Raman spectra. Pebax was treated with PIII of 20 keV nitrogen ions with 10^{17} ions/cm^2 fluence (Forschungszentrum Rossendorf, Dresden, Germany). Micro-Raman spectra were recorded in a backscattering mode excited by Nd:YAG laser irradiation (2ω, $\lambda = 532.14$ nm) on a diffraction double monochromator spectrometer HR800, Jobin Yvon with LabRam System 010. Optical microscope was used for focusing the exciting laser beam and for collection of Raman scattered light.

Detailed analysis of the carbonized layer structure by Raman spectra can be performed based on a three-stage model of Raman spectra depending on carbon structure [2]. For analysis we have to consider carbon atoms in different hybridization states. The electron configuration of free carbon atoms is the $1s^2 2s^2 2p^2$ state. In molecules the electron orbits of carbon exist in the following hybridization form: sp or sp^2 or sp^3 depending on the kind of molecule. Due to different hybridization the carbon has different allotrope forms: diamond-like (sp^3) and graphite-like (sp^2). Due to different ordering and orientation of carbon atoms in different hybridization states, clustering and hydrogen content, the carbon structures are classified as follows: diamond, diamond-like carbon DLC, amorphous carbon a-C and hydrogenized amorphous carbon a-C:H, tetrahedral amorphous carbon ta-C and hydrogenized tetrahedral amorphous carbon ta-C:H, graphite, fullerite (fullerene-containing),

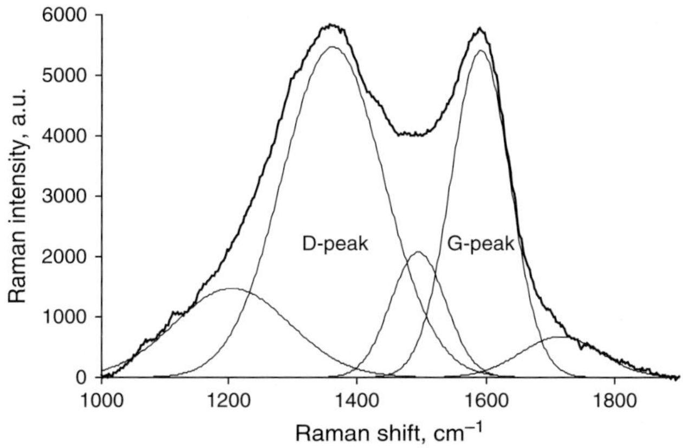

Fig. 4.15. Fitting result of Pebax Micro-Raman spectra. Pebax was treated with PIII of 30 keV nitrogen ions with 10^{17} ions/cm^2 fluence (Forschungszentrum Rossendorf, Dresden, Germany). Micro-Raman spectra were recorded in a backscattering mode excited by Nd:YAG laser irradiation (2ω, $\lambda = 532.14$ nm) on a diffraction double monochromator spectrometer HR800, Jobin Yvon with LabRam System 010. Optical microscope was used for focusing the exciting laser beam and for collection of Raman scattered light.

carbon nanotubes and nanofoam, chaoite and carben. The transformation of the target polymer under ion beam implanted is random. There is no clear evidence to suggest that the ions are directed to certain parts of the highly ordered carbon structure than others. Hence there is a stochastic character to the target structure transformation. Mostly, carbon in the nanocrystalline graphite (sp^2) and amorphous (sp^3) form as well as different mixtures of these two forms with different ordering result from ion beam implantation of the polymer surface layer.

The phenomenological model (proposed by Ferrari and Robertson [2]) is useful for analysis of carbon structure by Raman spectra. The analysis is based on G-peak position and the I(D)/I(G) ratio of D-peak and G-peak intensities. Real carbon structure created by deposition or ion beam implantation is considered as the stage of amorphization of well-ordered graphite structure (sp^2) up to the formation of tetrahedral amorphous carbon (sp^3). Four different forms of carbon are considered:

1. graphite (well-ordered infinite graphite plane);
2. nanocrystalline graphite (graphite sp^2 clusters with limited size are separated by edge defects, but sp^2 hybridization of carbon remains);
3. amorphous carbon a-C (graphite sp^2 clusters with smaller cluster size and about 20% of sp^3 bonds);
4. tetrahedral amorphous carbon (sp^3 hybridization bonds dominant and form DLC structures with sp^2 hybridization bond inclusions).

The Raman spectra of carbon have a resonance character and analysis must be carried out using a specific wavelength of excitation laser light. Let us consider Raman spectra excited by 514 nm laser line. Raman spectra of well-ordered graphite are characterized by the G-peak position at 1580 cm^{-1} and I(D)/I(G) = 0. With disordering of graphite structure, the G-peak shifts to the 1600 cm^{-1} position and I(D)/I(G) increases to 2. It corresponds to the nano-graphite (nc-graphite) structure. The L_a characteristic size of graphite clusters can be calculated by Tuinstra and Koenig's formula [3]:

$$\frac{I(D)}{I(G)} = \frac{C(\lambda)}{L_a} \tag{4.7}$$

where $C(\lambda)$ is constant depending on laser excitation wavelength. For a 515.5 nm wavelength $C(\lambda)$ is about 4.4 nm [2]. This formula works up to 2 nm of L_a. Following disordering converts the edge defects into a sufficiently significant amount of sp^3 bonds; however, the graphite structure is still a major fraction (up to 80%). The G-peak shifts to 1510 cm^{-1} and the I(D)/I(G) ratio decreases to about 0.2. The position of the G-peak depends on excitation laser line. The L_a characteristic size of graphite clusters becomes proportional to the I(D)/I(G) ratio:

$$\frac{I(D)}{I(G)} = C'(\lambda) \cdot L_a^2 \tag{4.8}$$

where $C'(\lambda)$ is about $0.55\,nm^{-2}$. This is a-C amorphous carbon. The graphite clusters are less than 2 nm.

With the degradation of graphite bonds, the number of sp^3 orbitals increases, such that ta-C tetrahedral amorphous carbon is formed. The position of the G-peak shifts from 1510 to $1570\,cm^{-1}$. The I(D)/I(G) ratio decreases to 0.

In polymers after ion beam implantation we have never seen Raman spectra correspond to well-ordered graphite or ta-C structures. Usually, the carbon layer has a structure between nc-graphite and amorphous carbon.

Following the present model, our Raman spectra of ion beam implanted polymers can be analyzed. Figure 4.14 shows Raman spectrum of Pebax after PIII with 20 keV energy nitrogen ions. The fitting of the spectrum by Gauss peaks gives narrow lines related to polyamide/polyether macromolecule vibrations and broad peaks of carbon structure at $1542\,cm^{-1}$ (G-peak) and $1332\,cm^{-1}$ (D-peak). The I(D)/I(G) ratio is 1.2. The G-peak position and I(D)/I(G) ratio correspond to carbon structures between nc-graphite and a-C states. Therefore, carbon cluster size can be calculated using formula (4.8). It gives an L_a of 1.5 nm.

Figure 4.15 shows Raman spectrum of Pebax after PIII with 30 keV energy nitrogen ions. The fitting of the spectrum by Gauss peaks gives broad peaks of carbon structure at $1590\,cm^{-1}$ (G-peak) and $1362\,cm^{-1}$ (D-peak). The I(D)/I(G) ratio is 1.74. The G-peak position and I(D)/I(G) ratio correspond to carbon structures between nc-graphite and a-C states. But the carbon structure of the present sample is closer to nc-graphite than the sample after PIII with 20 keV ions energy. Carbon cluster size is calculated by formula (4.8). It gives an L_a of 1.8 nm.

The fitting of Raman spectra of polyethylene after PIII shows that the nc-graphite structure is observed with 2 nm size of graphite clusters.

The carbonization effect is observed also by the presence of free electrons in graphite-like structures by electron paramagnetic resonance or electron spin resonance (EPR or ESR) spectra. The EPR spectra of polyethylene after PIII by

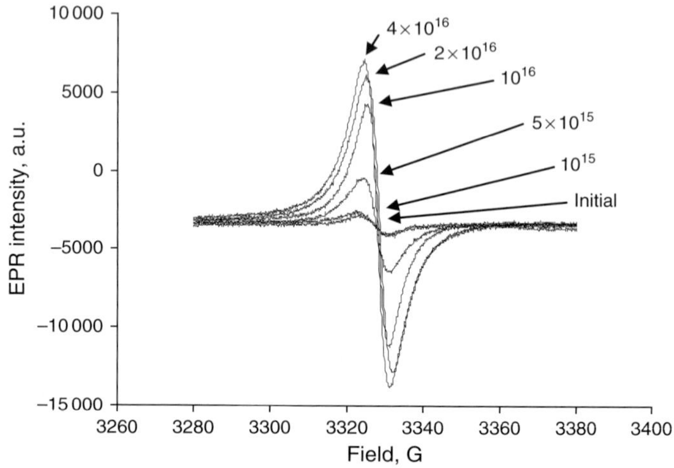

Fig. 4.16. Electron paramagnetic resonance (EPR) spectra of polyethylene film (LDPE, 50 μm) after PIII of 20 keV nitrogen ions (University of Sydney, Australia). Fluence of ion beam implantation is in ions/cm^2 units. EPR spectra were recorded by K. Fisher.

nitrogen ions of 20 keV energy are presented in Fig. 4.16. The symmetrical peak is observed with a g-factor of 2.0028. This value is close to the g-factor found in not-coupled electron EPR spectra of amorphous carbon structures (2.0027), which are formed at carbon deposition techniques and at pyrolis of polymers at high temperature in vacuum or in inert atmosphere. The signal intensity corresponds to the amount of free uncoupled electrons of carbon structures in ion beam implanted polymer. For example, the integral intensity of the EPR signal increases with fluence of PIII presented in Fig. 4.17. The signal becomes visible after 10^{15} ions/cm^2 fluence and increases asymptotically. At high fluence (5×10^{16} ions/cm^2 and higher) the signal intensity reaches saturation point, which shows saturation of carbonization process in the surface polymer layer. At the next fluence increase, the amount of carbon structure with free uncoupled electrons does not increase. This EPR signal is stable over a period of a few days where the sample is storing under laboratory conditions (25°C, atmospheric pressure of air).

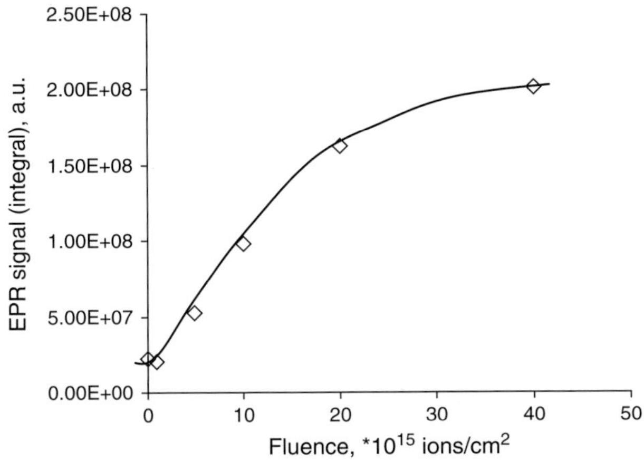

Fig. 4.17. Integral intensity of EPR spectra of polyethylene dependent on PIII fluence of treatment.

It means that such free electrons in the carbonized polyethylene surface layer do not disappear with time after ion beam implantation. The carbon structure is sufficiently stable.

The effect of structure transformations in polymers surface after ion beam implantation could be observed by Fourier-transform Infrared Attenuated Total Reflection (FTIR ATR) spectra. Such kind of spectra shows vibration modes of molecules. The intensity of the spectral lines is proportional to derivative of dipole moment of molecule on vibrational coordinate. Therefore, polar fragments of macromolecules are well observed in FTIR spectra.

For attenuated total reflection (ATR) effect in FTIR spectra the polymer sample is placed on an ATR prism crystal with a high refractive index in the infrared region. The crystal must be transparent in the infrared region. Usually, crystals of Ge ($n = 4.37$), KRS-5 or TlBr/TlI mixture ($n = 2.45$), KRS-6 ZnSe ($n = 2.49$) are used for the middle infrared region (4000–700 cm^{-1}). Detailed information about FTIR ATR spectroscopy can be found in [4]. The sample is pressed to the crystal to enable good optical contact between crystal and polymer. Due to this reason, soft or elastic polymers as well as thin films of polymer are preferable for

ATR spectra. The infrared beam penetrates the ATR crystal into the surface layer of the sample and returns to the crystal due to the effect of total reflection, when light crosses the boundary between optically dense media to optically soft media at an angle higher than the critical angle for such media. Therefore, the FTIR ATR spectra give information about the thin surface layer of the sample. The thickness of infrared beam penetration can be calculated using the following equation:

$$d_p = \frac{\lambda}{2\pi \cdot \sqrt{(n_0^2 \cdot \sin^2\alpha - n_1^2)}} \tag{4.9}$$

where λ is the wavelength of the infrared beam, α is the angle of crystal prism (incident angle of infrared beam), n_0 is the refractive indexes of crystal and n_1 is the refractive indexes of the polymer sample. Usually, the thickness of the polymer analyzed layer is 0.2–0.6 μm for Ge crystal and 0.8–2.5 μm for KRS-5 crystal in the middle IR region (3000–1000 cm^{-1}). Therefore, the analyzed layer is usually thicker than the modified polymer layer (about 0.01–0.1 μm) after ion beam implantation with middle ion energies. The FTIR ATR spectrum shows the vibrational modes of macromolecules in the bulk unmodified polymer layer and in the thin modified layer.

The example of FTIR ATR spectra of polyethylene ion beam implantation is presented in Fig. 4.18. All spectra show strong lines of polyethylene macromolecule vibrations at 2920 cm^{-1} and 2850 cm^{-1} of C-H stretch vibration modes, 1462 cm^{-1} and 1375 cm^{-1} of CH_2 and CH_3 deformation vibration modes, and 720 cm^{-1} and 730 cm^{-1} doublet of out-of-plane CH_2 vibration modes. With fluence increase, the intensity of all spectra lines becomes lower due to the formation of a carbonized layer. The carbonized layer is hard and optical contact of the polyethylene sample with the ATR crystal of the spectrometer worsens. However, the bulk unmodified layer of polymer becomes far from ATR crystal and its contribution in the spectra becomes

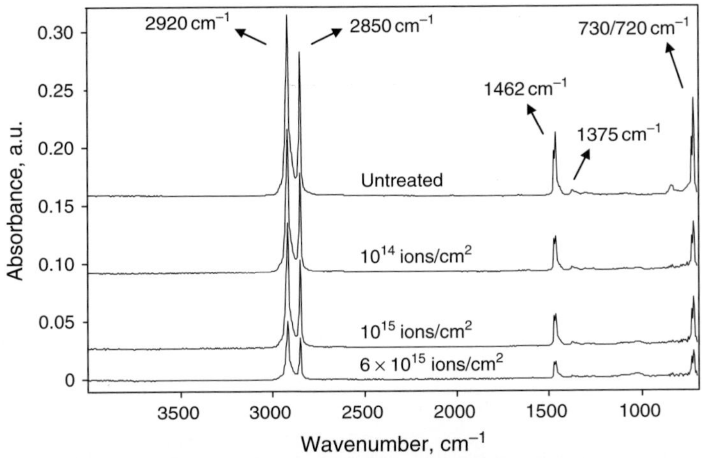

Fig. 4.18. FTIR ATR spectra of polyethylene (LDPE) film after ion beam implantation by nitrogen ions with 20 keV energy in pulse regime, pulse duration 0.3 ms, pulse frequency 1 Hz. Ion beam source is "Pulsar" in the Institute of Technical Chemistry, Perm, Russia.

less at high fluence of ion beam implantation. The spectral changes are weak and cannot be recognized without expanding the absorbance scale. The basic vibrational lines of polyethylene macromolecule become broader at high fluence (10^{16} ions/cm^2) of ion beam implantation because of the collection of structure defects.

For quantitative analysis the FTIR ATR spectra must be normalized with respect to the intensity of the main lines of the unmodified polymer (Fig. 4.19). The normalized spectra show changes which are related to the structure of polymer after modification.

In the spectra of treated polyethylene the lines of the carbonyl group at 1600–1750 cm^{-1} appear as a result of the oxidation process of the surface after ion beam implantation. The oxidation is caused by the reaction of the active surface layer with atmospheric oxygen after removing the modified sample from the vacuum chamber of the implanter. If the sample is not exposed on atmospheric oxygen, the oxidation effect is not observed. We did not see any oxygen-containing group lines in the 1600–1750 cm^{-1}

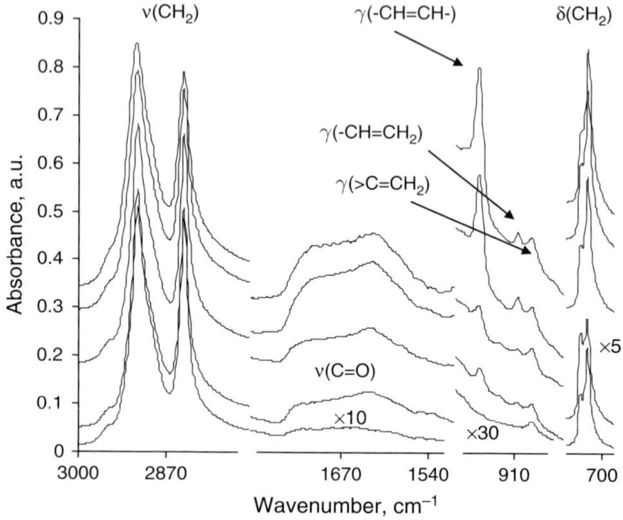

Fig. 4.19. Normalized FTIR ATR spectra of polyethylene after 20 keV nitrogen ions PIII with fluence (from bottom): initial, 10^{13}, 5×10^{14}, 5×10^{15}, 10^{16} ions/cm^2. Intensity of spectra is normalized on the 2913 cm^{-1} line. For a good view the multiplication factors for different spectra regions are applied.

region of FTIR ATR spectra of polyethylene after PIII, when spectra were recorded inside the vacuum chamber without polyethylene samples contacting with the atmosphere after PIII.

The band of the carbonyl group vibrations is very broad and contains some individual peaks. The fitting procedure and deconvolution of the carbonyl band could give a number of narrow peaks, which could be attributed to carbonyl bond vibration in different kinds of oxygen-containing groups. Some of the lines could be interpreted based on both the well-known spectra of individual substances as well as on the spectra of the surface after selecting reactions with separate oxygen-containing groups. For example, lines at 1685 and 1692 cm^{-1} could be attributed to unsaturated ketones; lines at 1706 cm^{-1} to the carboxylic acid group; lines at 1726 cm^{-1} to saturated ketones; lines at 1740 cm^{-1} to aldehydes; and lines at 1754 and 1765 cm^{-1} to esters.

The same lines are presented in the spectra of polyethylene after UV light irradiation. Some of the lines (for example, 1712 and 1718 cm^{-1} and other high frequency lines) could be attributed to carbonyl groups on edges of the graphitic or aromatic ring structures.

The shape of the carbonyl group band is unchanged after washing the modified polyethylene surface with organic inactive solvents like pentane and toluene. This means that the oxidized layer is associated with the polymer network and cannot be removed from the polymer. It was found that the shape of the carbonyl line did not change for polyethylene treated with nitrogen, oxygen or argon ions up to a fluence of 10^{16} ions/cm^2. This seems reasonable since the concentration of implanted ions in the treated polymer surface layer is significantly less than the concentration of polymer target atoms. The implanted ions play the role of a bomb, but not in the implanting phase.

For example, the FTIR ATR spectra of polyethylene after ion beam implantation by vacuum oil ions are presented on Fig. 4.20. The polyethylene film was treated with a very

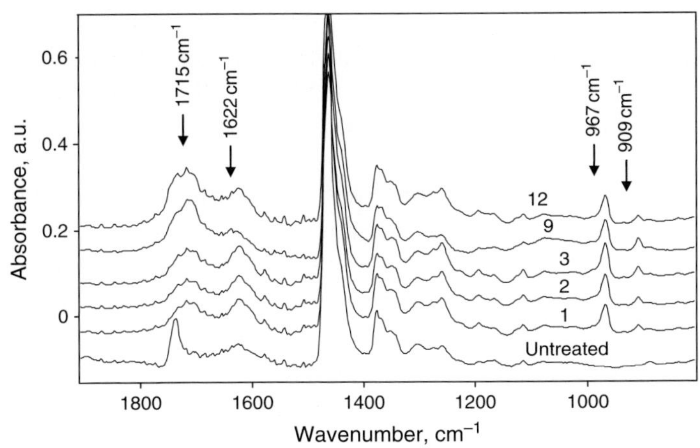

Fig. 4.20. FTIR ATR spectra of polyethylene treated with ion beam of vacuum oil ions with 300 keV energy on a TEMP accelerator (Institute of Nuclear Physics, Tomsk, Russia). Pulse duration is 50 ns, current density is 60 A/cm^2. Number of pulses are noted on graph.

high current density beam (60 A/cm^2). However, the pulse was short enough (50 ns) and the average current density was quite low (1 μA/cm^2) for excluding the overheating of the polyethylene film. The spectrum of initial polyethylene shows ordinary lines of polyethylene macromolecules and additionally a weak peak at 1737 cm^{-1} related to antioxidant in the film. After ion beam implantation, the wide band in the 1600–1750 cm^{-1} region of the spectra appears as a result of surface layer oxidation. There are no additional lines of vacuum oil molecules or their fragments in the FTIR ATR spectra.

The same region (1620–1670 cm^{-1}) of the spectra can be attributed to C=N vibrations in the FTIR spectra. Sometimes weak intensity lines to indicate C=N and C-N groups are observed by XPS spectra too. Compared to the previously discussed lines, the C=N and C-N lines are more difficult to interpret because molecular nitrogen is less active than oxygen and the reaction of molecular nitrogen with free radicals of polymer surface layer is less probable; also the extinction coefficient of the ν(C=N) lines is less than for carbonyl group vibration. Therefore, even with the clear presence of nitrogen-containing groups in the surface layer of the polymer after ion beam implantation, the contribution of ν(C=N) line intensity to the 1600–1750 band of FTIR spectra is expected to be less than the carbonyl line intensity.

In the same 1600–1650 cm^{-1} region the lines of unsaturated carbon-carbon structures may be observed and attributed to vibrations in C=C bonds. Such lines have quite low intensity in the infrared spectra and are overlapped with carbonyl group lines. More intensive lines attributed to out-of-plane C—H vibrations in unsaturated groups are observed in the 800–1000 cm^{-1} region:

887 cm^{-1}	in vinyliden group	$>C{=}CH_2$
906 cm^{-1}	in vinyl group	$-CH{=}CH_2$
964 cm^{-1}	in vinylen group	$-CH{=}CH-$

These lines are situated in the transparent region of the FTIR spectra of polyethylene and are clearly observed after ion beam implantation.

FTIR spectra can be used for quantitative analysis of structure changes in the polymer surface layer after ion beam implantation. The analysis is based on the Beer–Lambert–Bouguer law:

$$Log_{10}\left(\frac{I_0}{I}\right) = \varepsilon \cdot C \cdot d = A \qquad (4.10)$$

where I_0 is the incident light intensity, I is the passed light intensity, d is thickness of absorbance sample, C is the concentration of absorbing groups, ε is extinction coefficient, and A is the absorbance value.

In FTIR ATR spectroscopy the thickness d has the meaning of effective propagation depth of infrared beam and depends on wavelength. Therefore FTIR ATR spectra have different absorbance in different spectral regions. However, usually the absorbance depends on the quality of optical contact between the ATR crystal and the sample. Such effects do not permit the use of the Beer–Lambert–Bouguer law directly for quantitative analysis of the spectra. However, the relative quantitative analysis of structure can be done using spectra normalization. For analysis, the absorbance (intensity) of spectral lines which are related to the vibrations of investigated structure (A_1) must be normalized with respect to the intensity of lines (A_2) which are related to the stable groups or structure (internal standard):

$$\frac{A_1}{A_2} = \frac{\varepsilon_1 \cdot C_1 \cdot d_1}{\varepsilon_2 \cdot C_2 \cdot d_2} = \frac{\varepsilon_1 \cdot C_1 \cdot \lambda_1}{\varepsilon_2 \cdot C_2 \cdot \lambda_2} \qquad (4.11)$$

where λ_1 and λ_2 are positions of lines 1 and 2. The presence of a wavenumber is due to different penetration depths of the infrared beam into the sample in ATR effect. Therefore, if the sample has gradient structure, as in the case of ion beam treated polymers, the spectral information comes from different layers at λ_1 and λ_2 wavelengths. Therefore, equation (4.11) indicates that the internal standard line position (λ_1) must be as close as possible to the position of the analyzed line (λ_2).

Following this requirement, the analysis for the carbonyl group line in the 1600–1750 cm^{-1} region, the 1462 cm^{-1} line of deformational vibrations in the polyethylene macromolecule can be used as the internal standard line (Fig. 4.21).

The analysis of carbonyl group intensity in polyethylene gives an asymptotical curve with an increase of fluence of ion beam implantation. The carbonyl group concentration increases at low fluence of treatment (up to 10^{15} ions/cm^2) and then it is saturated. Fluence increase does not lead to a rise in oxidation product concentration. The same asymptotical character of concentration with fluence is observed for vinyl groups. However, the vinylen groups' concentration continues to grow with fluence of ion beam implantation and the intensity of the vinylen groups is higher than the

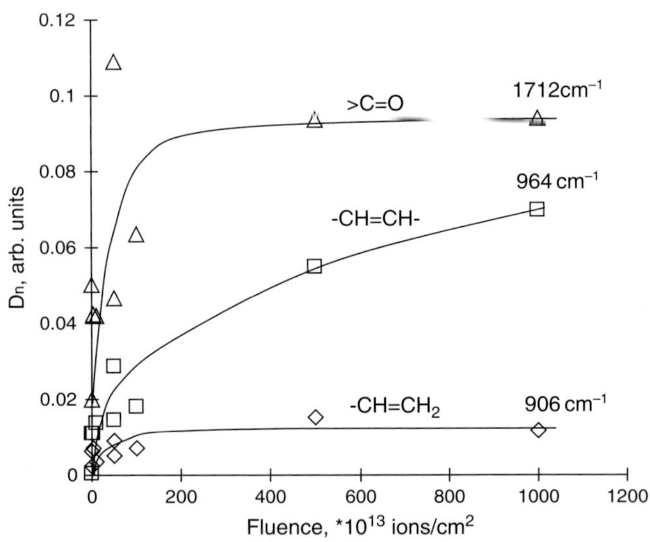

Fig. 4.21. Optical density (absorbance) of ATR FTIR spectra lines of polyethylene (LDPE) on fluence of PIII by nitrogen ions of 20 keV energy (Forschungszentrum Rossendorf, Dresden, Germany).

intensity of the vinyl and vinyliden groups. The highest intensity of the vinylen groups is interpreted as the main product in the destruction process of polyethylene macromolecules in free radical reactions. Due to the high carbonization stage of polyethylene in the surface layer, the vinylen groups cannot survive in the thin layer with high structure defects. However, the vinylen groups can survive and be collected in deeper layers than ion penetration depth. The free radical reactions of free valence travel along the macromolecules can lead to vinylen groups in layers much deeper than ion penetration depth.

Similar spectra changes are observed in the FTIR ATR spectra of other olefine polymers after ion beam implantation. For example, the FTIR ATR spectra of EPDM rubber after PIII by nitrogen ions are presented in Fig. 4.22. The strong lines at 1462 and 1375 cm^{-1} are shown oversaturated. The weak lines at 1630, 1597, 1503, 1153 and 945 cm^{-1} are related to vibrations in the dicyclopentadiene fraction of the EPDM macromolecule. In spite of this, a very weak broad

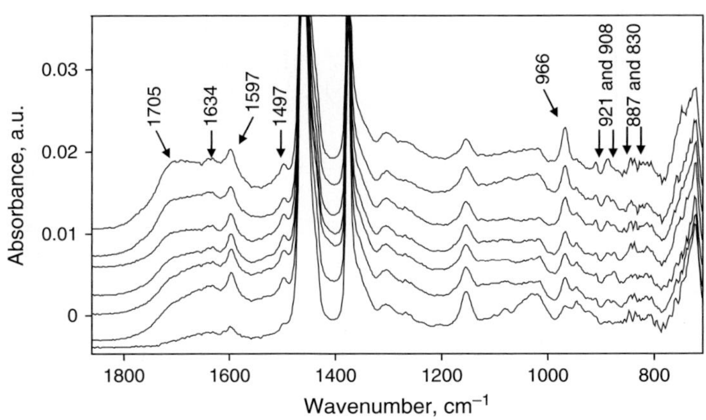

Fig. 4.22. FTIR ATR spectra of EPDM rubber without vulcanizing agents and fillers after PIII by nitrogen ions with 20 keV energy (Forschungszentrum Rossendorf, Dresden, Germany). Fluence of ion beam implantation from bottom to top is as follows: untreated, 10^{13} ions/cm^2, 10^{14} ions/cm^2, 5×10^{14} ions/cm^2, 10^{15} ions/cm^2, 5×10^{15} ions/cm^2, 10^{16} ions/cm^2. Position of lines is noted in cm^{-1}.

band at the 1650–1750 cm^{-1} region is related to a surface oxidized layer under environmental conditions.

After ion beam implantation the new broad band at 1600–1750 cm^{-1} appears as the result of surface layer oxidation; the narrow lines at 1597 and 1497 cm^{-1} show a presence of unsaturated cycles which are synthesized under ion beam; the line at 966 cm^{-1} of the vinylen groups, the lines at 921 and 908 cm^{-1} of the vinyl groups and the lines at 890 and 830 cm^{-1} of vinyliden groups show the presence of new unsaturated C=C groups in irradiated rubber as the result of ion beam implantation.

The FTIR ATR spectra of ethylene and octane-1 co-polymer are presented in Fig. 4.23. The spectrum of untreated polymer shows strong lines of aliphatic hydrocarbon vibrations of CH, CH_2 and CH_3 groups at 2920, 2850, 1464, 1377 and 720 cm^{-1}. The weak narrow line at 1725 cm^{-1} corresponds to vibration of the carbonyl group in the surface layer oxidized under environmental conditions. After ion beam implantation a broad line appeared

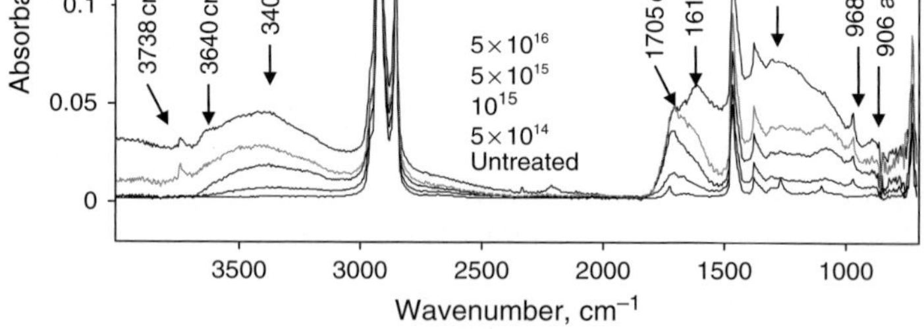

Fig. 4.23. FTIR ATR spectra of co-polymer of ethylene and octene-1 (EG-8200) after PIII by 30 keV energy nitrogen ions (Forschungszentrum Rossendorf, Dresden, Germany). Fluence of ion beam implantation from bottom to top spectra (ions/cm^2) is noted.

in the 1600–1750 cm^{-1} region corresponding to vibration of C=O and C=C groups. The new lines at 968, 906 and 888 cm^{-1} corresponding to CH out-of-plane vibrations in unsaturated C=C groups also appear. The very broad lines in the 3200–3700 cm^{-1} region correspond to hydroxyl group vibrations. Separate maxima of this line at 3400 and 3500 cm^{-1} correspond to vibrations of the hydroxyl groups which are connected by hydrogen bonds. The narrow peak at 3738 cm^{-1} is caused by vibrations of the free hydroxyl group. With fluence increase, the intensity of oxygen-containing group lines increases. At high fluence of ion beam implantation the very broad band with a maximum at 1250 cm^{-1} is observed, which corresponds to carbon—carbon vibrational states in the carbonized surface layer of polymer. Such a band is observed in polyethylene and other polymers under high fluence of implantation (higher than 5×10^{16} ions/cm^2).

The FTIR ATR spectra of butyl rubber after ion beam implantation show fewer intensive changes than polyethylene and its co-polymers (Fig. 4.24). The spectrum of untreated butyl rubber shows main vibrational peaks at 1230, 1366, 1389, 1471, 2895, 2950 and 2975 cm^{-1}. After modification by ion beam, the spectra show a weak broad band at 1600–1730 cm^{-1} corresponding to C=O and C=C groups and a 3300 cm^{-1} broad band corresponding to OH group vibrations. However, the spectra of treated rubber show lines at 2920 and 2850 cm^{-1}, which are attributed to vibrations of the CH_2 group. Here we have to remind ourselves that the butyl rubber is a decomposed polymer under different kinds of irradiation. The depolymerization of butyl rubber occurs under UV, electron and g-irradiations. The reaction of depolymerization in butyl rubber is considered in Chapter 3. However, depolymerization is not a unique reaction which occurs in butyl rubber under ion beam. Due to high density defects and high temperature in the ion trace, the reactions of free radicals result the new groups of hydrocarbons, which are synthesized by the rearrangement of separated groups and atoms and the formation of simple

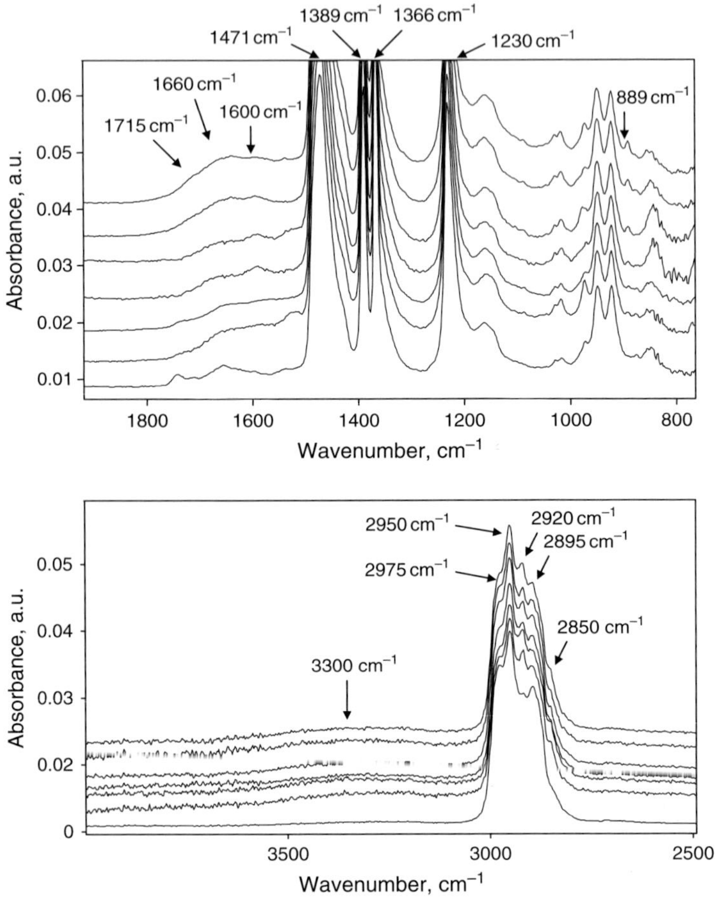

Fig. 4.24. FTIR ATR spectra of butyl rubber after PIII by nitrogen ions of 20 keV energy (Forschungszentrum Rossendorf, Dresden, Germany). Fluence of treatment is as follows (from bottom to top): initial, 10^{13}, 10^{14}, 5×10^{14}, 10^{15}, 5×10^{15}, 10^{16}.

aliphatic structures such as —CH< and $—CH_2—$ groups. As a result, strong lines of CH_2 vibrations appear in the spectra. With increase in fluence of treatment, the intensity of the lines which represent such groups increases.

The presence of out-of-plane CH vibrations in the 967 and 910 cm^{-1} regions is covered by two intensive lines at 923 and 950 cm^{-1} of butyl rubber. A weak peak at 889 cm^{-1}

corresponding to the vinyliden groups is observed in spectra of high fluence treated samples.

The spectra which reflect structural changes of polymers with heteroatom in the backbone, modified by ion beam are more difficult to analyze. For example, the FTIR ATR spectrum of polyethylene terephthalate (PET) contains a number of high intensity lines at 1717, 1342, 1260, 1124, 1104 and 1021 cm^{-1} which are attributed to vibration modes of highly polar groups: C—O and C=O (Fig. 4.25). The lines of aromatic ring at 1581 and 1507 cm^{-1} and lines of the aliphatic group at 2925, 2854 and 1472 cm^{-1} have very low intensity. After ion beam implantation, the spectra change slightly: absorbance in the 1600–1700 cm^{-1} and 1400–1000 cm^{-1} regions becomes higher. The absorbance becomes higher with fluence increase. Such weak spectral changes are well observed in differential spectra of PET before and after ion beam implantation. The differential spectrum shows clearly a new broad peak with a center at 1617 cm^{-1}, corresponding to carbonyl and C=C groups in the surface layer of

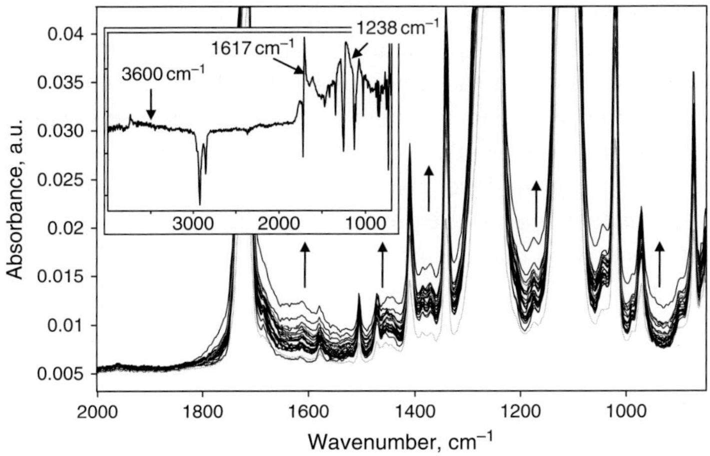

Fig. 4.25. FTIR ATR spectra of polyethylene terephthalate (PET) after PIII by nitrogen ions with 20 keV energy (University of Sydney, Australia). Array shows spectral trend for fluence increase from untreated to 10^{16} ions/cm^2. Inserted is the differential spectrum: 10^{16} ions/cm^2 treated sample spectra minus untreated sample spectra.

PET after ion beam implantation. Also, at high fluence of treatment, a strong broad peak with a center at $1238\,cm^{-1}$ attributed to carbon structure is observed. In the region of OH vibration, a weak broad peak with a center at $3600\,cm^{-1}$ related to hydrogen bonded hydroxyl groups and a narrow peak at $3745\,cm^{-1}$ related to free hydroxyl groups are observed. These spectral changes as well as the decrease of the 1250 and $1125\,cm^{-1}$ lines' intensity show that the effect of carbonization of PET surface layer under ion beam is combined with degradation of the ester group and formation of hydroxyl groups.

PET is one of the examples of degradable polymers under irradiation. Together with oxidation and carbonization effects, the chemical decomposition under ion beam is observed in polymers with heteroatoms in the backbone. In the case of polyethers, amides, imides and similar more complex structure polymers than polyethylene, the effects of hydrolysis, regrouping, and substitution reactions are evident.

Another example of degradation reactions under ion beam implantation is polyurethane. There are different kinds of polyurethane which depend on the structure of the macromolecule. One of the kinds of polyurethane is called segmented polyurethanes, which have microphase segregation of the structure. FTIR ATR spectra of polyurethane based on polyoxypropylene glycol, polyoxyethylene glycol and toluenediisocyanate are presented in Fig. 4.26. The structure of this polyurethane is as follows:

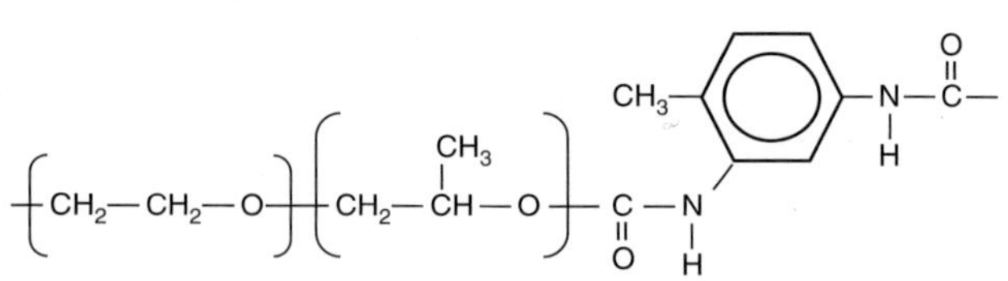

(4.1)

The spectra contain strong lines at 1102, 1225, 1536, 1725, 2873, 2974 and $3327\,cm^{-1}$. These lines correspond to vibrational modes of mostly polar groups in polyurethane

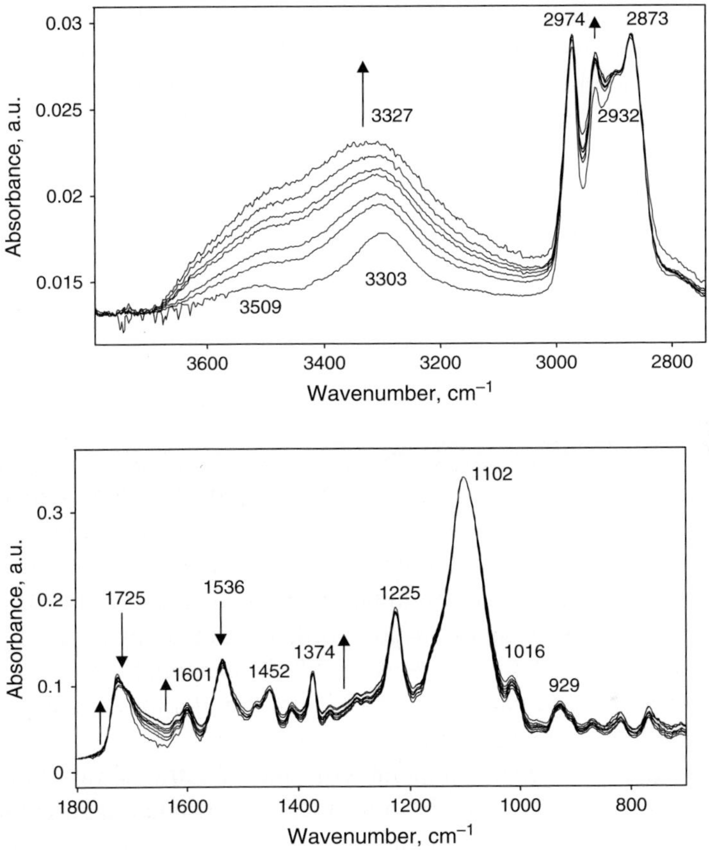

Fig. 4.26. FTIR ATR spectra of polyurethane based on polypropylene glycol terminated by toluene diisocyanate, and polyethylene glycol (Laprol L 2080) after PIII by nitrogen ions with 20 keV energy (Forschungszentrum Rossendorf, Dresden, Germany). The treatment fluence is as follows (by array): initial, 10^{15}, 1.3×10^{15}, 10^{16}, 1.3×10^{16}, 2×10^{16}, 5×10^{16}, 10^{17} ions/cm^2. Polyurethane was synthesized by K. Kondyurin.

macromolecules. These are urethane (—NH—CO—O—) and ether (—C—O—) group vibrations. The urethane group has well-identified characteristic lines: Amide A (3327 cm^{-1}), Amide 1 (near 1725 cm^{-1}) and Amide 2 (1536 cm^{-1}), which can be used for the structure analysis of the urethane group. The presented polyurethane has two parts of macromolecule, which are not compatible with each other: polyether and urethane fragments. The presence of two incompatible fragments causes a microphase separation in polyurethane. The first phase is formed by polyether fragments of macromolecules. This phase is amorphous and has high flexibility. The second phase is formed by urethane groups connected with each other by hydrogen bonds:

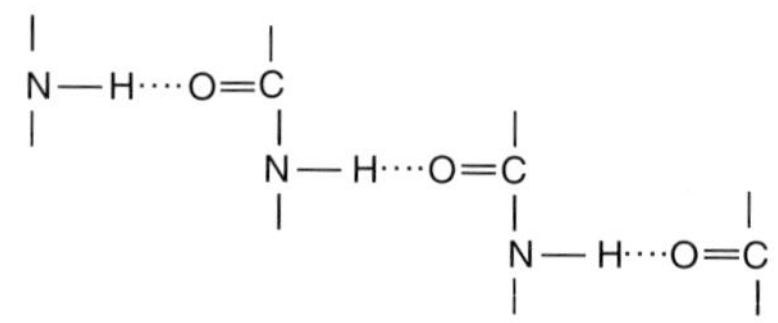

(4.2)

Such a phase has semi-crystalline hard structures called domains. The presence of hard domains in polyurethane is very important. It gives perfect mechanical properties of polyurethane in combination with high elasticity. The presence of domains in polyurethane can be analyzed by form of the Amide 1 peak in FTIR spectra. In our case, the peak has a maximum at 1725 cm^{-1} and a low frequency shoulder up to 1650 cm^{-1}.

After ion beam implantation the spectra show some changes similar to olefin molecules. In particular, additional absorbance appears in the 1600–1700 cm^{-1} region as an increase in broad band intensity. This band is caused by carbonization and oxidation of the carbon structure. The intensity of such a line is sufficiently weaker than is observed in the spectra of olefin polymers after ion beam implantation. A small decrease in Amide 1 and Amide 2 line intensity is observed with fluence of ion beam implantation. At the same fluence, the sufficiently strong increase of absorbance in the 3600–3300 cm^{-1} region is observed as the appearance of a number of hydroxyl group vibrations.

All such spectral changes correspond to the complex hydrolysis process of polyurethane macromolecules. These chemical transformations are caused by energy transfers and excitation of separate groups of macromolecules which take part in decomposition reactions of irradiated polymer in regions deeper than ion penetration depth. In such regions the free radical diffusion occurs and causes chemical processes similar to UV irradiations and γ-irradiations. As a result, a much deeper layer of polymer could be activated by implanted ions especially in polymers that are not stable under any kind of irradiation.

A highly degradable polymer under ion beam is polytetrafluorethylene (PTFE). This polymer has a very low carbonization effect. The FTIR ATR spectra show small changes which are slightly higher than the level of noise (Fig. 4.27). The broad band with two maxima at 1714 and 1651 cm^{-1} of the oxidized carbon structure as well as the hydroxyl group line at 3400 cm^{-1} (only at high fluence of treatment) are observed after ion beam implantation. The intensity of these lines increases slightly with fluence of ion beam implantation.

The PTFE surface becomes very attractive to any pollution from the surrounding atmosphere. Sometimes, the FTIR ATR spectra show a high concentration of hydrocarbons of different structure (and FTIR spectra) in the surface layer

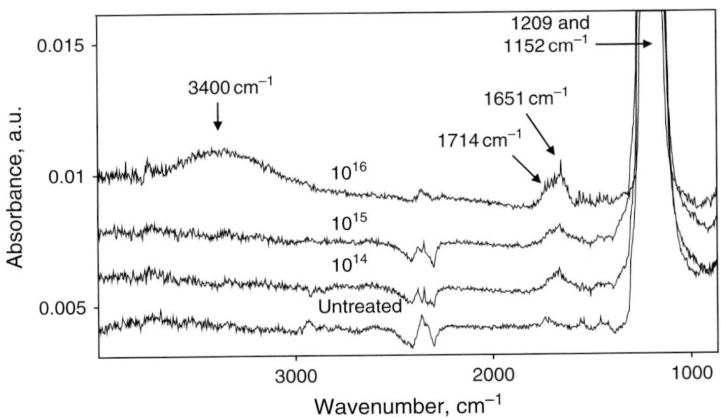

Fig. 4.27. FTIR ATR spectra of polytetrafuorethylene (PTFE, Teflon) after PIII of nitrogen ions with 20 keV energy (Leibniz-Institut für Oberflächenmodifizierung e.V., Leipzig, Germany). Fluence of treatment is noted in ions/cm^2.

after ion beam implantation. However, careful treatment of PTFE before and after implantation results in an absence of any hydrocarbons on the top layer of the PTFE after ion beam treatment.

The structure transformations are collected with an increase of ion beam fluence. The effects of structure changes can be calculated quantitatively by FTIR spectra. Figure 4.28 shows values for relative concentrations of carbonyl and unsaturated carbon groups and hydroxyl groups determined by FTIR ATR spectra in polyurethane with fluence of ion beam implantation. The lines at 1660 and 3300 cm^{-1} are used for calculations by equations (4.10) and (4.11). The concentration of carbonyl and unsaturated carbon group increases at low fluence and has saturation at high fluence (more than 10^{16} ions/cm^2). A similar character of dependence is observed for the concentration of hydroxyl groups. Hydrolysis can proceed in a field of polymer where the macromolecules of the polymer are not yet carbonized,

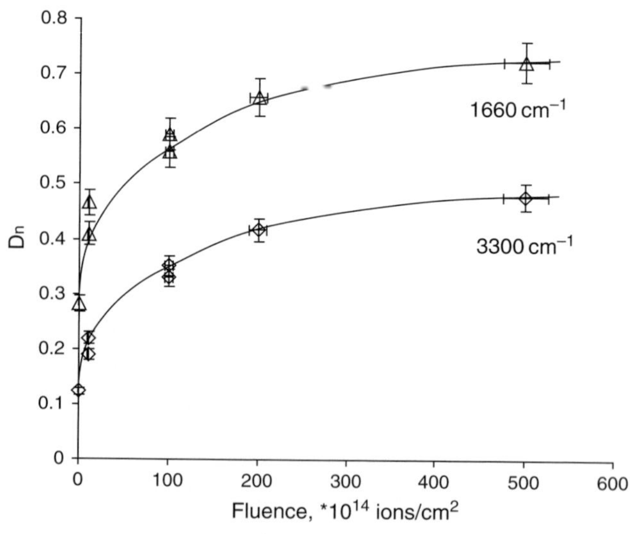

Fig. 4.28. The normalized absorbance of 1660 and 3300 cm^{-1} lines in FTIR ATR spectra of polyurethane with fluence of ion beam implantation.

but are activated by radiation effects of ion penetration. In these regions the macromolecules accept diffused oxygen from the atmosphere to form oxygen-containing groups and to take part in the hydrolysis process with depolymerization of macromolecules and formation of glycols.

If the polymer layer becomes highly carbonized, oxidation does not occur, because of low activity of amorphous carbon reacting with molecular oxygen at normal conditions. The effect of inactive carbonized polymer to oxidation is observed for the thin layer of polystyrene on silicon wafer (Fig. 4.29). When polystyrene film is treated with ion beam with low fluence, the modified layer is thin and the oxidation effect increases with fluence increasing (picture a). When whole polystyrene film is completely modified and the carbonization of the film becomes significant, the following ions cannot produce a higher amount of active polystyrene macromolecules (picture b). When the polystyrene film becomes carbonized, the activity of the film to oxygen decreases despite the

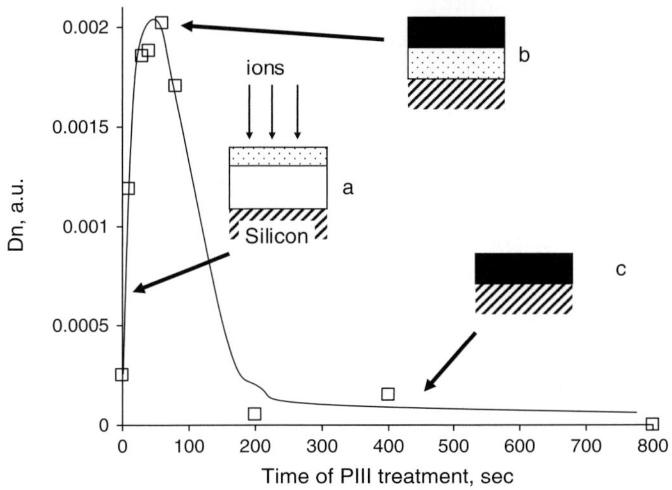

Fig. 4.29. Carbonyl group line intensity in FTIR transmission spectra of polystyrene layer (100 nm thickness) after PIII.

sufficiently thick residual carbonized film on the silicon wafer (picture c).

In comparison with plasma treatment without bias, ion beam treatment has a longer period of oxidation kinetics under normal atmospheric conditions. This is caused by stabilization of the surface layer of the polymer in the case of ion beam implantation, because crosslinked and highly carbonized structures appear in the surface layer of the polymer. The penetration depth of ion plays a significant role in stabilization. The modified layer must be thick enough to exclude the free movement of molecules in the polymer surface layer. The required thickness depends on polymer structure and the kind of polymer macromolecules. The effects of globular structure (thickness must be more than one globule in diameter), etching rate (to create enough carbon fraction) and fillers (absorbers of ion energy) are important for prediction of the kinetics of structure transformations in the polymer surface layer after ion beam implantation. The post-history kinetics of ion beam irradiated polymers has not yet been investigated in detail. However, some experimental data could be used to predict the structure and properties evolution during ageing of the modified polymer surface.

For example, the active environment also has an influence on relaxation processes after ion beam implantation. The FTIR ATR spectra of polyethylene (UHMWPE) with time after PIII treatment are presented in Fig. 4.30 for high fluence (4×10^{16} ions/cm^2) and in Fig. 4.31 for low fluence (5×10^{14} ions/cm^2). In both spectra the lines of the oxygen-containing groups in the 1600–1750 cm^{-1} and 3000–3600 cm^{-1} regions appear "immediately" after treatment (about 3 minutes after switching off the pulses and removing the sample from the vacuum chamber on ATR crystal). We could not record FTIR ATR spectra any more quickly after stopping ion beam implantation; however, the FTIR ATR spectra of the polyethylene sample after ion beam implantation without contact with air oxygen do not show any carbonyl or hydroxyl groups lines. In the spectra

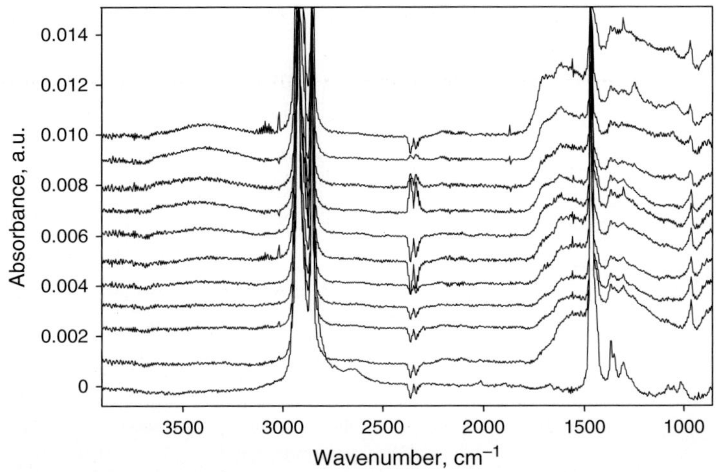

Fig. 4.30. FTIR ATR spectra of ultrahigh molecular weight polyethylene (UHMWPE) after PIII by nitrogen ions of 20 keV energy with fluence of 4×10^{16} ions/cm^2. The spectra are recorded with time of storing under room conditions (air, 25°C, 50–60% of moisture) after stopping of ion beam implantation (from bottom to top): untreated, 3 min, 33 min, 61 min, 126 min, 220 min, 1 day 2 hours, 4 days, 5 days 4 hours, 7 days 2 hours, vent in air 3 weeks after. The experiment was done in cooperation with B.K. Gan.

of the high fluence treated sample, the broad band from 1000 cm^{-1} to 1700 cm^{-1} is observed, which corresponds to carbon structures of the surface layer.

The lines of unsaturated carbon—carbon groups (vinyl, vinylene and vinyliden) do not change with storage time for the treated polyethylene after PIII with high and low fluence. The 1712 cm^{-1} line of carbonyl groups and the 3500 cm^{-1} line of hydroxyl groups become more intensive with storage time after treatment. The dependence of carbonyl line intensity on ageing time after PIII is presented in Fig. 4.32. The saturated concentration of carbonyl groups is achieved more quickly for low fluence of PIII. For high fluence, the saturation was not even achieved during the duration of the experiment. The initial concentration of the carbonyl group determined immediately after PIII does not

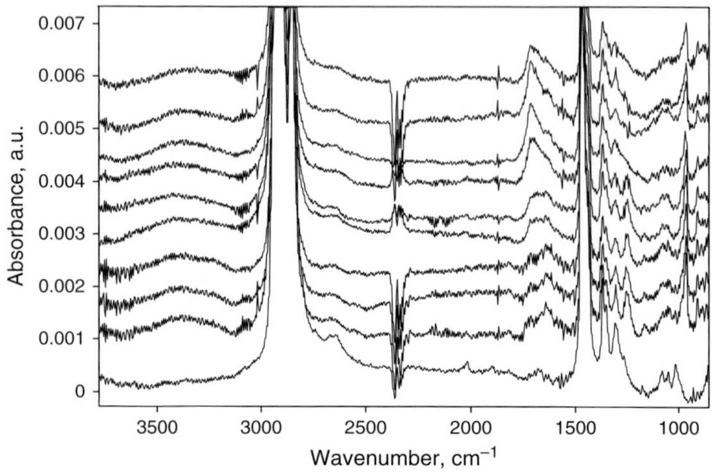

Fig. 4.31. FTIR ATR spectra of ultrahigh molecular weight polyethylene (UHMWPE) after PIII by nitrogen ions of 20 keV energy with fluence of 5×10^{14} ions/cm^2. The spectra are recorded with time of storing under room conditions (air, 25°C, 50–60% of moisture) after stopping of ion beam implantation (from bottom to top): untreated, 3 min, 31 min, 63 min, 123 min, 187 min, 4 days 10 hours, 5 days 4 hours, 7 days 2 hours, vent in air 3 weeks after. The experiment was done in cooperation with B.K. Gan.

show a dependence on treatment fluence. The difference appears after some storage time. Therefore, the concentration of oxygen-containing groups grows continually for high fluence implanted samples, when the oxidation process of low fluence implanted samples is stopped. The saturation concentration is higher for higher fluence of PIII. As reported in Chapter 3, the free radical reactions with atmospheric oxygen are responsible for the oxidation kinetics of ion beam implanted polymer. The difference in carbonyl group accumulation for different fluences shows the different reaction kinetics of free radicals with oxygen.

A similar time-dependent effect is observed for the surface energy of the polymer. The wettability of the polymer surface changes sufficiently strongly after ion beam implantation. For example, the contact angle of

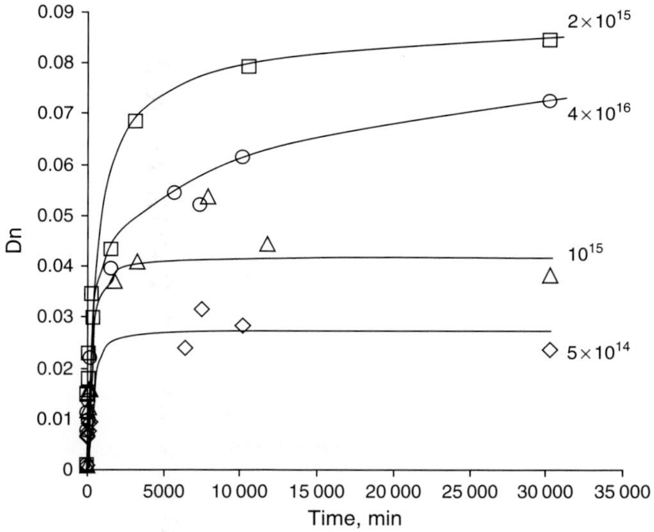

Fig. 4.32. Normalized optical density (absorbance) of $\nu(C{=}O) = 1712\,cm^{-1}$ line in ATR FTIR spectra of polyethylene (UHMWPE) with time of storing after PIII by 20 keV nitrogen ions. Fluence of treatment (ions/cm^2) is noted on the graph.

polyethylene with water changes from 90 degrees for untreated to 15–20 degrees for ion beam treated polymer. The changes are observed at low and high fluence of treatment. For example, for polyethylene it starts to be seen from 10^{12} to 10^{13} ions/cm^2 fluences. The wettability is an experimental characteristic of the surface energy of polymer, which depends on structure and composition of the thin surface layer. The increase in wettability after ion beam implantation shows that the surface layer becomes hydrophilic and more polar chemical groups appear in this layer. Immediately after implantation the wettability is extremely high. With storage time of the polymer after ion beam implantation, the wettability becomes lower as it is observed by the wetting angle of a water drop (Fig. 4.33). The kinetics of wettability depends on applied fluence of ion beam implantation. For low fluence of implantation, the wetting angle reverses more quickly than for the high fluence treated

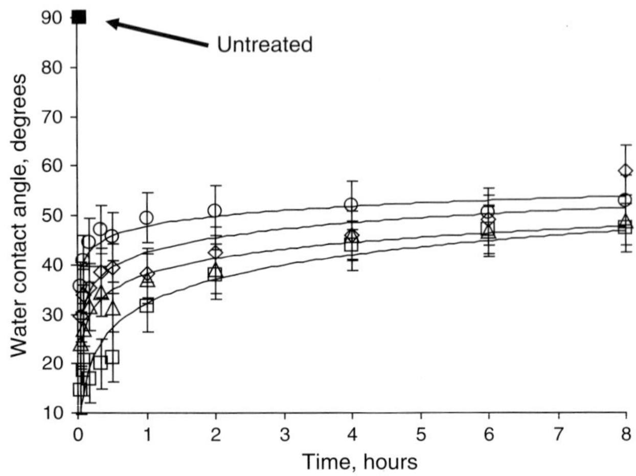

Fig. 4.33. Wetting angle of water drop on UHMWPE surface treated with 20 keV energy nitrogen ions with PIII fluence of 5×10^{14} (circle), 10^{15} (rhomb), 5×10^{15} (triangle), 4×10^{16} (cubic) ions/cm^2 with storing time under room conditions. Contact angle measurements were done by B.K. Gan.

sample. The stable wetting angle of ion beam modified polymers dependent on fluence of implantation varies by 50–70 degrees.

The changes in wettability are caused by ageing of the modified polymer surface with time after ion beam implantation. However, the kinetics of wettability is quicker than spectral transformation at the same parameters of treatment and storing conditions. The kinetics of wettability is associated with the collapse of free radicals and changes in surface structure due to environmental influence, for example oxidation. The oxidation process has a complex character; it depends on the type of polymer, structure of the surface layer, temperature and environmental contamination. It also supports the complex character of free radical transformations and oxygen-containing group formation during exposure to air. In spite of this, the carbon structure of the polymer surface layer also has an influence on wettability. Therefore, the wettability of the polymer surface after ion beam implantation is a result of free radicals, oxygen-containing groups and carbonization of the polymer surface layer. Variation of free radical concentration and oxygen-containing group concentration is a reason for the changes in wettability at ageing of the modified polymer surface. Detailed consideration of the wettability of polymers after ion beam implantation is presented in Chapter 5.

The effect of structure changes in the surface layer after PIII is observed using X-ray photoelectron spectra (XPS) of treated polymers. For XPS the polymer sample is irradiated by X-ray photons and the electrons from atoms of the top layer of macromolecules are spattered and analyzed. The kinetic energy of spattered electrons equals the energy of photons minus bonding energy of electrons in the atom. Therefore, XPS gives a value of electron energy levels in the atom that depends on bonds with neighboring atoms. In XPS the thin layer of polymer (up to 1 nm) is analyzed. For example, in spectra of initial polyethylene the single C_{1s} peak related to carbon bonded with two carbon atoms and two hydrogen atoms is observed at 285.0 eV. In spectra of ion beam implanted polyethylene the main peak at 285.0 eV has satellites at 286.5, 288.0 and 289.4 eV which are attributed to the carbon atom in oxygen-containing groups (Fig. 4.34). The line at 286.5 eV is related to carbon in the C—O group, the line at 288.0 eV is related to carbon in the carbonyl C═O group and the line at 289.4 eV is related to carbon in the O—C═O group. The ratio of intensities of these lines corresponds to the ratio of their concentrations.

Also the O_{1s} line appears in the XPS of ion beam implanted polyethylene as a result of reactions with air oxygen. The line is fitted with two individual peaks at 531.3 and 535.8 eV, which are attributed to the oxygen atom in the C═O and C—O groups, respectively.

Despite ion beam implantation by nitrogen ions, the line of N_{1s} is not presented in the 400.3 eV region of the spectrum. The fluence of treatment (10^{16} ions/cm^2) is not enough for the implanting nitrogen ions to bind to the polyethylene surface and to be detected by the XPS spectrometer.

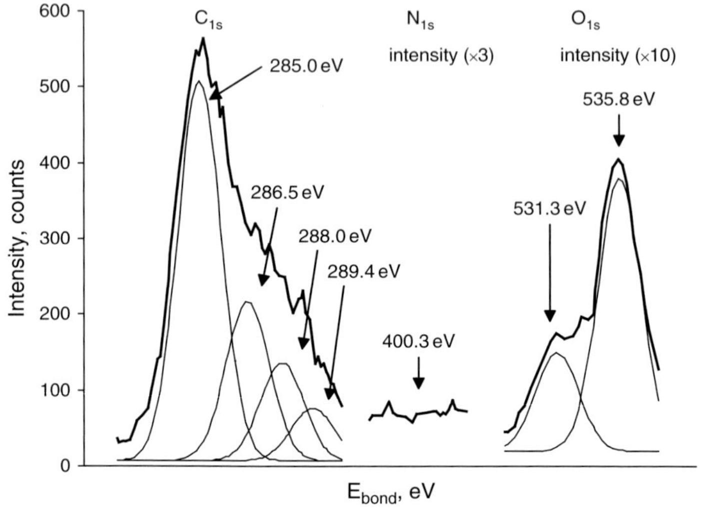

Fig. 4.34. X-ray photoelectron spectra (XPS) of polyethylene (LDPE) after ion beam implantation by nitrogen ions with 20 keV energy and 10^{16} ions/cm^2 fluence. The polyethylene was treated on a Pulsar ion beam source (Institute of Technical Chemistry, Perm, Russia). The spectra were recorded on XPS spectrometer with magnetic separation of electrons in the laboratory of Prof. V.A. Trapeznikov (Physical-technical Institute, Izhevsk, Russia).

For such spectra we used a spectrometer with magnetic separation in an electron analyzer.

Another XPS spectrometer gave us the spectra of polyisoprene rubber with higher signal/noise level and the N_{1s} peak is clearly observed in the spectrum of ion beam implanted polymer (Fig. 4.35). The intensity of the nitrogen line is much smaller than the intensity of the O_{1s} line. Even implantation by nitrogen does not give a significant amount of nitrogen in comparison with the oxidation processes in polymer after exposure to air. It supports a model when implanting ions play the role of bombs penetrating the polymer network and causing breaks of bonds. The effect of implantation with a collection of implanting atoms starts at higher fluences, when the structure of the polymer is completely converted into inorganic carbon or silicon structure.

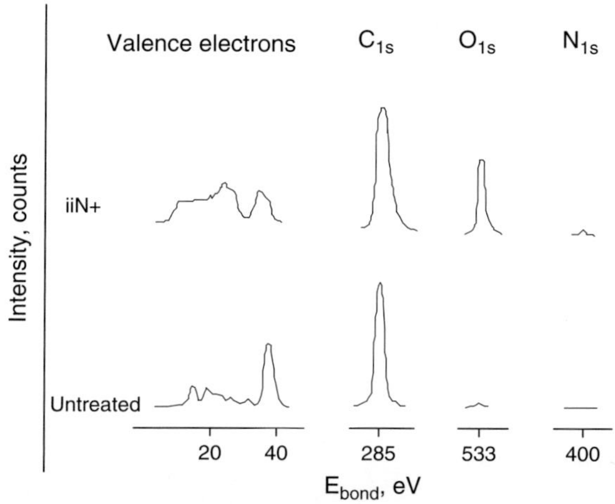

Fig. 4.35. XPS of polyisoprene (SKI-3NT) before and after ion beam implantation by nitrogen ions with 20 keV energy and 10^{16} ions/cm^2 fluence. Polyisoprene was treated on a Pulsar ion beam source (Institute of Technical Chemistry, Perm, Russia). The spectra were recorded by Olga Kanunnikova (Physical-technical Institute, Izhevsk, Russia).

Some polymers are very sensitive to X-ray irradiation during spectra recording. We could not reach a low enough intensity of X-ray beam for the PTFE sample because of low sensitivity of our detector at the time of experiment. Modern spectrometers have higher sensitivity and the effect of polymer degradation can be excluded. The spectrum of untreated PTFE shows a symmetrical F_{1s} peak at 690.2 eV, C_{1s} peak at 292.0 eV, and a number of low energy peaks of valence electrons with definite maximums at 37.0, 30.0, 15.0 and 6.0 eV (Fig. 4.36). Due to degradation under X-ray, the C_{1s} peak is observed at 285.9 eV and it grows with time of X-ray irradiation during spectra recording.

After ion beam implantation, the XPS of PTFE shows an F_{1s} peak at 690.2 eV, O_{1s} peak at 533.6 eV, N_{1s} peak at 400.6 eV, and C_{1s} peak at 285.9 eV. The peaks of F_{1s} and C_{1s} have complex shapes corresponding to the overlapping of some narrow peaks. These individual peaks support the presence of oxygen-containing and nitrogen-containing

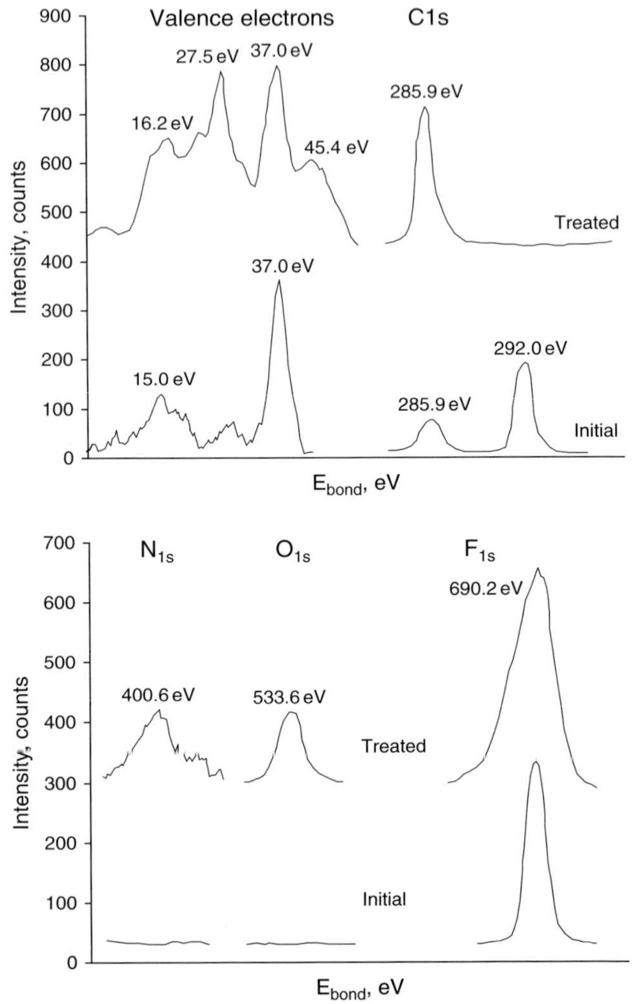

Fig. 4.36. XPS of polytetrafluorethylene (PTFE) before and after ion beam implantation by nitrogen ions with 20 keV energy and 10^{16} ions/cm^2 fluence. Intensity scales are different for different regions of the spectra for a good view of the shape of the lines. PTFE was treated on a Pulsar ion beam source (Institute of Technical Chemistry, Perm, Russia). The spectra were recorded by Olga Kanunnikova (Physical-technical Institute, Izhevsk, Russia).

groups in the surface layer of PTFE after ion beam implantation. The spectrum in a region of valence electrons is greatly altered. An interpretation of the valence electron spectrum is more complicated and it is not usually used for structure analysis of the polymer after ion beam implantation. However, this region of spectra gives more detailed information on valence electron structure and chemical structure of the irradiated layer of the polymer.

The quantitative analysis of element contamination can be done by XPS without details of chemical group analysis. The intensity of the XPS line is proportional to the concentration of elements with sensitivity factor. Using intensity of the lines the dynamics of element concentration is observed with fluence of ion beam implantation. For example, Fig. 4.37 shows a dependence of element contamination determined by integral intensity of C_{1s}, F_{1s}, O_{1s} and N_{1s} XPS lines on fluence of ion beam implantation by nitrogen ions with 20 keV energy. The thin surface layer analyzed by XPS (about 1 nm thickness) is defluorinated under low

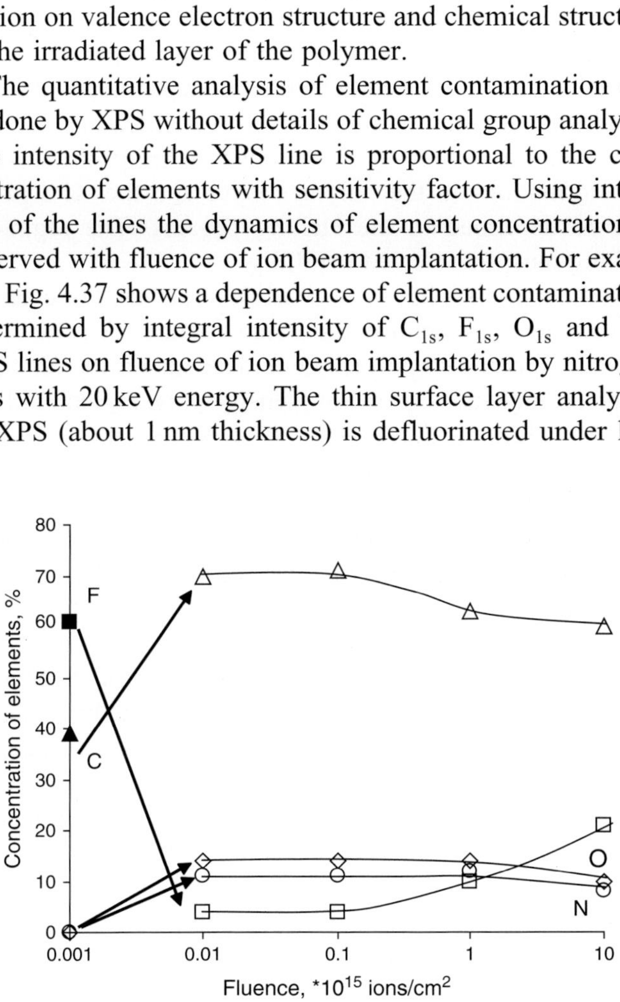

Fig. 4.37. Concentration of fluor, carbon, oxygen and nitrogen elements in PTFE after ion beam implantation with 20 keV nitrogen ions by XPS data.

fluence (10^{13} ions/cm^2). At the same time, the small amount of nitrogen and oxygen atoms are bonded to the activated PTFE surface. Following treatment does not increase the defluorination effect. At high fluence, the fluor concentration increases slightly, perhaps due to the etching effect.

Using the etching of the polymer layer in low energy argon beam and following XPS recording, the profile of element contamination can be built and compared with the defect profile. A good agreement of both profiles is observed for different kinds of polymers and penetrating ions [5]. However, the etching by argon ions also brings the structure transformations in the polymer layer and the XPS profile can be deformed in the etching process. The regime of etching must be selected carefully.

Some polymers have a crystalline structure. For example, polyethylene has an orthorhombic structure. The fraction of crystalline phase in polyethylene depends on the regularity of macromolecules. For low density polyethylene (LDPE) the crystalline fraction is about 40%. X-ray data shows clear diffraction patterns of untreated LDPE. A decrease in the crystalline phase of the polymer is observed with fluence of ion beam implantation. X-ray peaks of the polyethylene crystalline phase become broader and weaker after ion beam implantation (Fig. 4.38). Decreasing the crystalline peaks' intensity corresponds to the disappearance of the crystalline phase, which is also observed by FTIR spectra of polyethylene. The doublet of 720/730 cm^{-1} spectrum lines characterizing the ratio of crystalline/amorphous phases does not change with fluence of ion beam implantation, but the absolute intensity of such doublet lines decreases with fluence of ion beam implantation due to degradation of the polyethylene macromolecular structure. The disappearance of the crystalline phase of the polymer under ion beam occurs due to the disappearance of the polymer structure and its transformation into the carbon phase. In connection with crystalline structure, the polarization measurements of the stretched polyethylene film by FTIR spectra show that the macromolecular orientation of the film does not change after

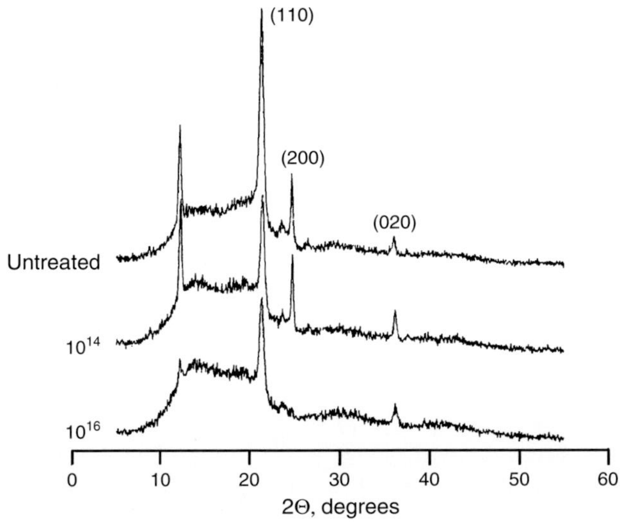

Fig. 4.38. XRD data of polyethylene (LDPE) before and after ion beam implantation with nitrogen ions of 30 keV energy (Institute of Electrophysics, Ekaterinburg, Russia). XRD data were recorded by V. Mizgulin.

ion beam treatment. Here we do not consider the thermal regime of implantation, where the polymer film or surface layer is heated up to melting point temperature or glass transition temperature (T_g). At thermal regime of implantation the effect of amorphization of the surface layer and disorientation of the film has been observed.

An important parameter of polymer structure is the topology of macromolecules. If a polymer has linear macromolecules, it can be dissolved in solution. If a polymer is treated by ion beam, it becomes a crosslinked insoluble fraction after treatment. The macromolecules become joined to each other by chemical bonds, which cannot be broken by intermolecular interactions of solvents. The polymer becomes one huge molecule. The crosslink formation is caused by free radical reactions at which the free radicals of neighboring macromolecules react and form a chemical bond (bridge) between the macromolecules. The crosslinking effect could be determined by gel-fraction measurements. For measurements, a solvent must be chosen so that the linear untreated polymer completely dissolves. The solvent must be chemically inert to the study polymer. After ion beam treatment, an insoluble fraction of polymer could not be dissolved in such solvent. The insoluble part of the polymer contains crosslinked macromolecules. If polymer is dissolved only in hot solvent, the polymer can be treated with boiling solvent in Sockslet apparatus. After separation from solution, the residual insoluble part of the polymer is dried and weighed. The gel-fraction content is determined as:

$$G = \frac{m_1}{m_0} \tag{4.12}$$

where m_1 is the weight of the insoluble part of the polymer, m_0 is the initial weight of the polymer before dissolving. The gel-fraction formation under any type of radiation could be described by the theory of Charleby–Pinner:

$$(s + \sqrt{s}) = \frac{p_0}{q_0} + \frac{10}{q_0 \cdot R \cdot u_1} \tag{4.13}$$

where s is the sol-fraction of the polymer, p_0 is the fracture density per unit dose in kGy, q_0 is the density of crosslinked units per unit dose in kGy, u_1 is a number of average degree of polymerization, and R is the radiation dose in kGy.

The crosslinking formation depends on the kind of polymer and it could be expressed as the gel-formation parameter for different kinds of polymers: the chemical yield $G(x)$ is a number of crosslinks per 100 eV deposited energy from ions. $G(x)$ is calculated from the assumption that the crosslinking process occurs proportionally to the linear energy transformation (dE/dx). In this case the chemical yield is calculated as:

$$G(x) = \frac{100 \cdot N_a}{2 \cdot M_{w,0} \cdot D_{gel}} \tag{4.14}$$

where $M_{w,0}$ is the initial average molecular weight of polymer macromolecules, D_{gel} is gel dose in eV/g, and N_a is

Avogadro's number. A similar parameter $G(s)$ could be introduced for the analysis of scissions in polymer under irradiation. $G(x)$ and $G(s)$ characterize the crosslinking and scission ability of polymer macromolecules under ion beam implantation. For example, the $G(x)/G(s)$ ratio for polyethylene is 1.2/0 and for PTFE is 0/(4–20), which means high crosslinking ability of polyethylene and high scission ability of Teflon under ion beam.

At such measurements it is assumed that the crosslinking process occurs uniformly in whole thicknesses of the polymer film or coating. If ions penetrate the film with negligible energy loss and the profile of the defect distribution in film thickness is close to constant value, the gel-fraction measurement gives a real value. It is observed for ions of very high energy and very thin polymer films. However, in most cases of industrial processes, the ion energies are not high enough and polymer films are not thin enough for the complete penetration of ions through the film. Therefore the gel-fraction estimation is not correct because the crosslinking profile caused by structure defects is not constant in whole thicknesses of the polymer film.

For applicable energy of ions and thickness of polymer film the structure defect profile drops down at some certain depth from the surface. In this case, the gel-fraction could be recalculated to the thickness of the crosslinked film:

$$h_g = \frac{m_1}{m_0} \cdot h_0 \tag{4.15}$$

where h_0 is the thickness of the polymer film before dissolving. At this calculation the density of the crosslinked part of the film is assumed to be equal to the density of the initial film and the crosslinked layer has a step profile on thickness.

For example, the dependence of the crosslinked fraction of polyethylene film treated with nitrogen ions with 20 keV energy is presented in Fig. 4.39. At low fluence the insoluble amount of polyethylene increases up to maximal value. The maximal thickness of the crosslinked layer is 1.75 μm which

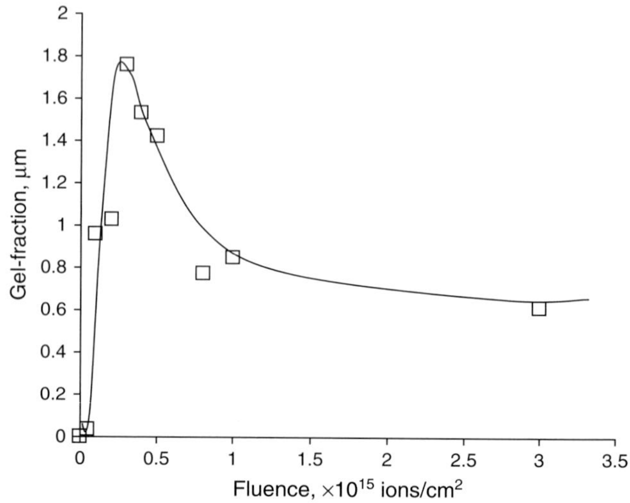

Fig. 4.39. Thickness of crosslinked layer (gel-fraction) of polyethylene (LDPE, 50 μm) after ion beam implantation with 20 keV energy nitrogen ions (Institute of Technical Chemistry, Perm, Russia). Gel-fraction measurements were done by E.N. Tereshatova.

is observed at a fluence of $3 \cdot 10^{14}$ ions/cm^2. It is more than 20 times the ion penetration depth (75 nm). At a higher fluence the crosslinked layer of polyethylene decreases up to 0.8 μm, it is ten times thicker than the ion penetration depth.

The color of the residual insoluble fraction of polyethylene is different and depends on fluence of ion beam implantation. At low fluence (10^{13} ions/cm^2) the color is white similar to untreated polyethylene. At higher fluence the color transforms in the following order: white-milk, cream-orange ($5 \cdot 10^{14}$ ions/cm^2), cream-brown (10^{15} ions/cm^2), red-green depending on angle of view ($3 \cdot 10^{15}$ ions/cm^2) and green-black (10^{16} ions/cm^2). The color changes correspond to absorbance in the UV-visual spectra of polyethylene as the conjugation number of unsaturated and aromatic structures in the modified surface layer of polyethylene increases.

A similar dependence of the crosslinked layer thickness is observed at implantation of oxygen (maximum is at 0.855 μm), carbon (maximum is at 0.84 μm) and argon (maximum is at 2.09 μ[μυρ]m) ions into polyethylene. Also similar dependence of gel-fraction is observed for polyisoprene and polystyrene films after ion beam implantation.

The formation of the thick crosslinked layer is explained by propagation of the free radicals due to chain reactions in the deep unmodified layer, diffusion of destruction products and penetration of secondary irradiations released at collisions of target atoms and implanting ions such as high energy phonons, electrons and photons into bulk layer deeper than ion penetration depth. Such secondary effects of the ion beam implantation are able to cut off hydrogen atoms in virgin macromolecules in bulk polymers and to generate crosslinks by free radical reactions.

The appearance of thick crosslinking layer in implanted polymer changes significantly properties of polymer thin films and coatings. The polymer coating becomes insoluble, even if the implanting ions do not fly through the polymer. For example, optical microphotos of polystyrene coating on silicon wafer and scanning electron microscope (SEM) images are presented in Figs 4.40 and 4.41. A polystyrene coating of 100 nm thickness was treated with argon ions with 20 keV energy. The ions penetrate into the 60 nm layer. Therefore, the carbonized layer can be no more than 60 nm thickness. Before treatment the polystyrene film can be washed with toluene flow for a few seconds. After treatment, the polystyrene film becomes insoluble even after low fluence of treatment (2.5×10^{14} ions/cm^2). There are some regions of polystyrene film which swell differently in toluene. After evaporation of toluene, the film is shrunken and wrinkled. With fluence increase, the wrinkling area becomes less and disappears at high fluence (10^{16} ions/cm^2). At highest fluence, the film is crosslinked sufficiently densely that the film cannot swell in toluene.

Due to crosslinking and carbonization of treated polymer, the compressed stresses appear in the surface layer. The stresses are strong enough to cause deformation in the thick surface layer of μm and tens of μm thickness.

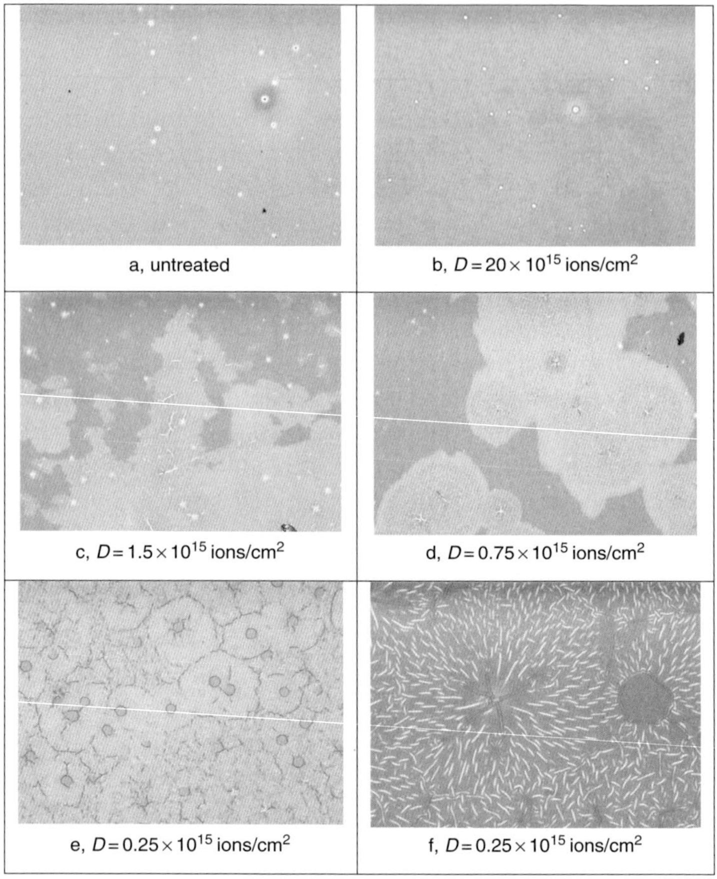

Fig. 4.40. Microphotos of spincoated polystyrene film (100 nm) on silicon wafer before (a) and after PIII (b–f) with different fluence of treatment. (b–f) films have been swelled in toluene and dried. Pictures (a–e) have 1 × 1.2 mm size, picture (f) has 20 × 24 μm size. The experiment was done in cooperation with K. Mizuno and B.K. Gan.

Such deformation is viewed as a bizarre morphology of polymer materials after ion beam implantation. The morphology of the surface depends on the type of polymer, regime of ion beam implantation, temperature, thickness of the film, fillers and other parameters. We observed structures such as parallel wrinkles in polyethylene (Figs 4.42

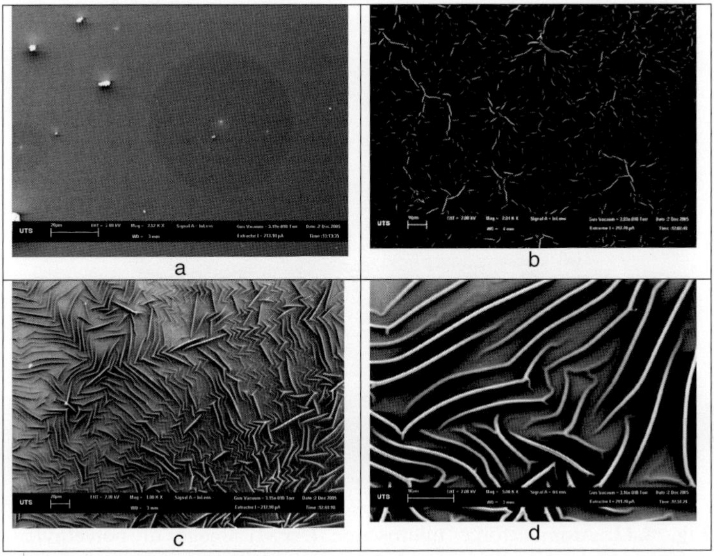

Fig. 4.41. Scanning electron microscope (SEM) images of spincoated polystyrene film on silicon wafer after washing in toluene. Before washing the films were treated with PIII: a – 20 keV energy and 2×10^{16} ions/cm^2 fluence; b – 20 keV energy and 0.75×10^{15} ions/cm^2 fluence; c and d – 5 keV energy and 1×10^{15} ions/cm^2 fluence of PIII (University of Sydney, Australia). SEM images were taken by R. Wuhrer. The experiment was done in cooperation with K. Mizuno and B.K. Gan.

and 4.43) and in silicon rubber (Fig. 4.44); "domain" structures of wrinkles in polyurethane (Fig. 4.45); stairs structures (Fig. 4.46), worm structures (Fig. 4.47) and stars structures (Fig. 4.48) in EPDM rubber; and fir tree-like structures in silicon rubber (Fig. 4.49). Such morphologies appear in the structures of two layers: the compressed hard surface layer on the top of soft layer. The soft layer can be unchanged or depolymerized under the surface layer. In both cases, the surface layer stresses are relaxed due to the deformation of the soft under layer. If the thin polymer film is stacked to a hard substrate (for example, silicon wafer or glass) and the soft undersurface layer is absent, the surface morphology does not change.

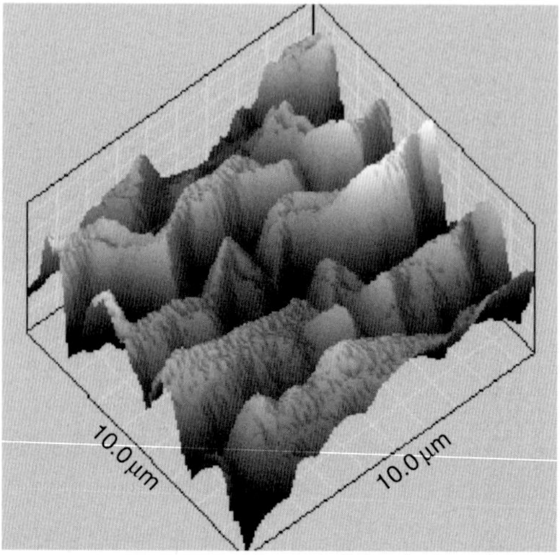

Fig. 4.42. Atomic force microscope (AFM) image of polyethylene (LDPE, 50 μm) surface after PIII with 20 keV nitrogen ions and 10^{16} ions/cm^2 fluence of treatment.

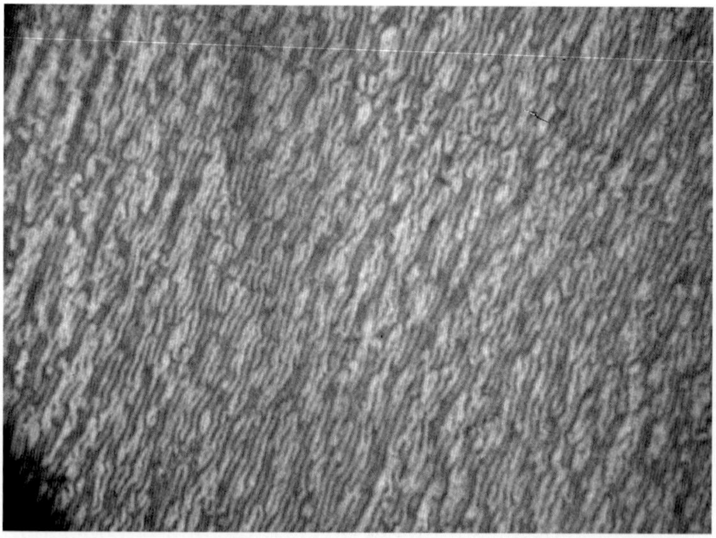

Fig. 4.43. Optical microphoto of polyethylene (LDPE, 50 μm) surface after PIII with 20 keV nitrogen ions and 10^{16} ions/cm^2 fluence of treatment.

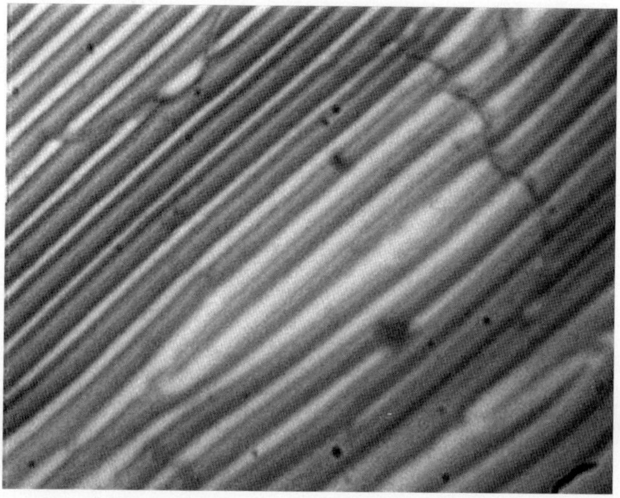

Fig. 4.44. Optical microphoto of silicon rubber after ion beam implantation with nitrogen ions of 20 keV energy.

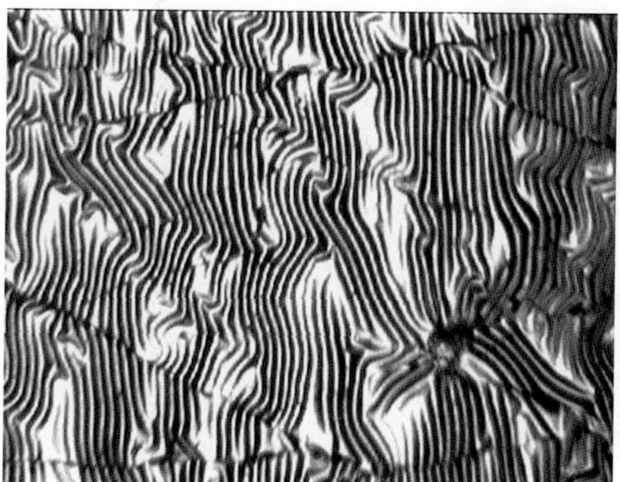

Fig. 4.45. Optical microphoto of polyurethane after ion beam implantation with 20 keV nitrogen ions.

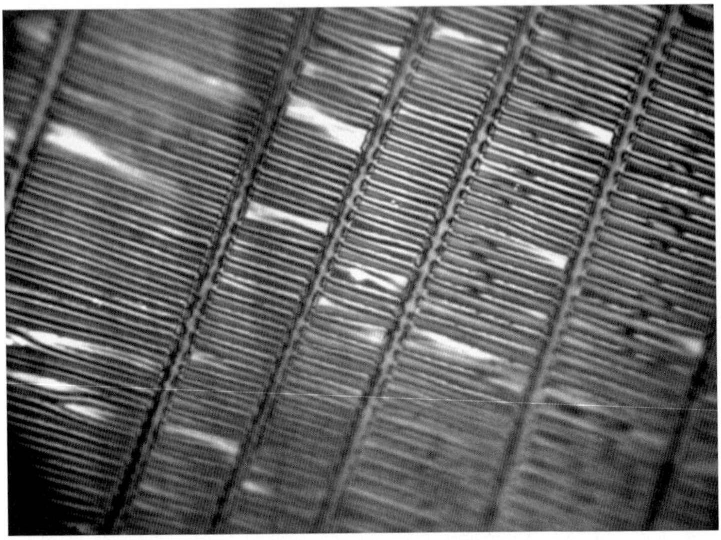

Fig. 4.46. Optical microphoto of EPDM-40 rubber surface after PIII with 20 keV nitrogen ions.

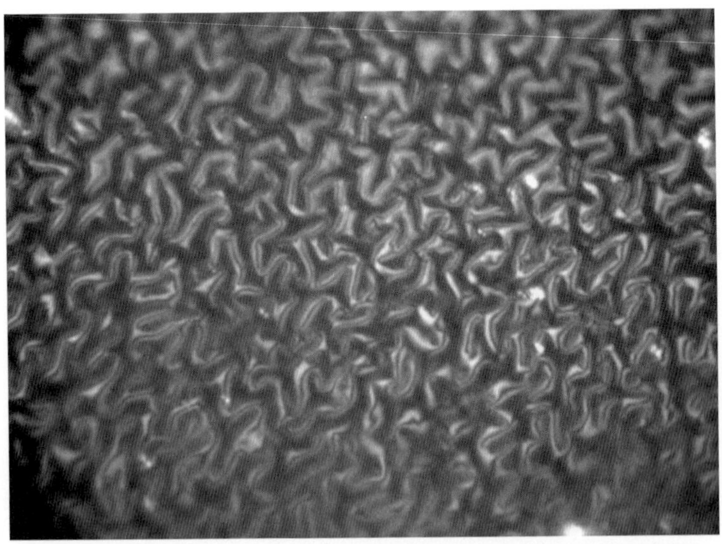

Fig. 4.47. Optical microphoto of EPDM-40 rubber surface after PIII with 20 keV nitrogen ions.

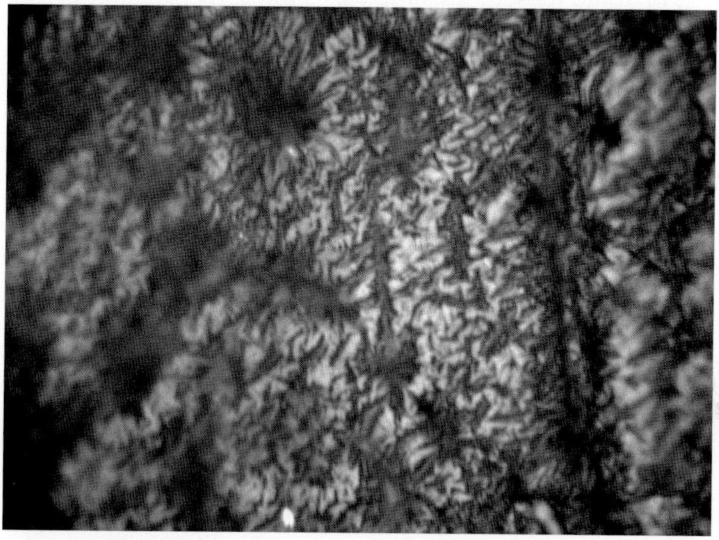

Fig. 4.48. Optical microphoto of EPDM-40 rubber surface after PIII with 20 keV nitrogen ions.

Fig. 4.49. Optical microphoto of silicon rubber after ion beam implantation with nitrogen ions of 20 keV energy.

For some reason, in some cases the compression stresses in the surface layer do not cause the deformation of the undersurface layer. In such cases, the stresses can cause cracks in the hard surface layer as seen in polyurethane (Figs 4.50 and 4.51) and butyl rubber (Fig. 4.52). The density and depth of cracks increase with fluence of treatment. In all cases, we observed the thickness of the hard layer was (depth of cracks) greater than ion penetration depth. Therefore, surface morphology changes are caused not only by the carbonized thin layer, but also by the crosslinked layer which is thicker than ion penetration depth.

The structure changes in polymer after ion beam implantation affect the mechanical properties of treated polymer. The mechanical behavior depends on the type of polymer, thickness of polymer film or coating and implantation parameters. As usual, the strength of polymer film decreases with fluence of treatment (Fig. 4.53). The modulus of elasticity increases due to carbonization and crosslinking, the elongation at break decreases due to brittleness increase of the polymer. It is observed for polymers with low

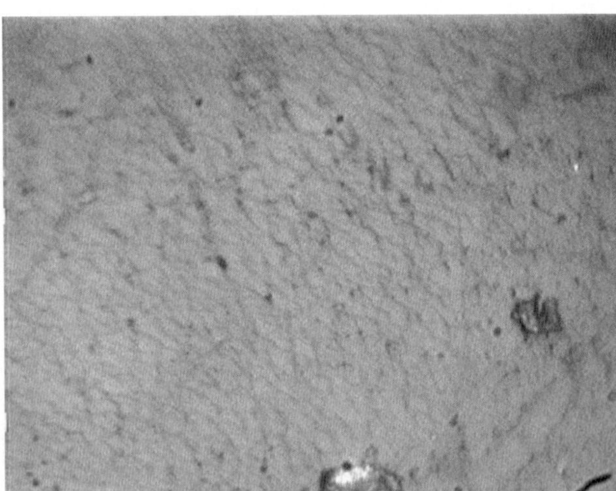

Fig. 4.50. Optical microphoto of polyurethane after ion beam implantation with 10^{14} ions/cm^2 fluence.

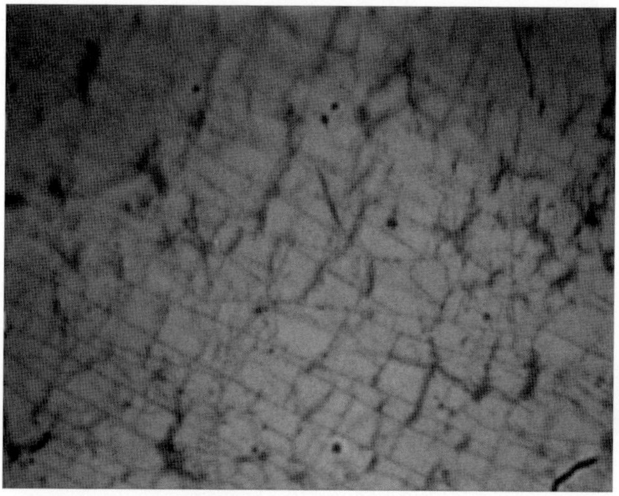

Fig. 4.51. Optical microphoto of polyurethane after ion beam implantation with 10^{16} ions/cm^2 fluence.

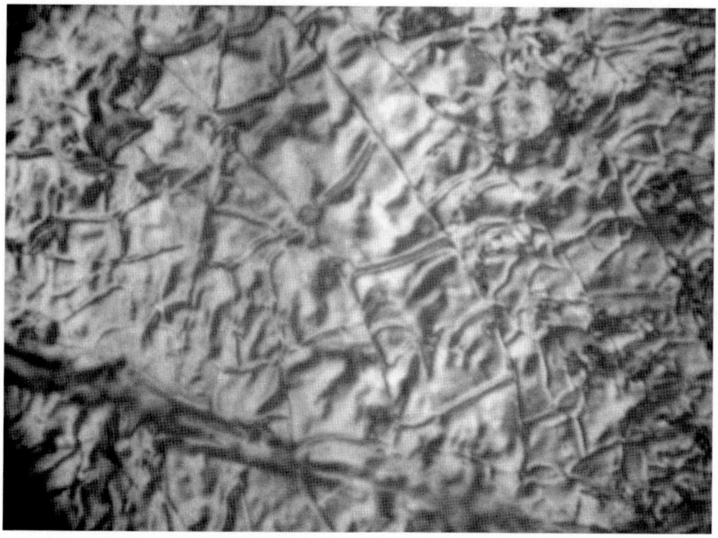

Fig. 4.52. Optical microphoto of butyl rubber surface after PIII with 20 keV nitrogen ions and 10^{16} ions/cm^2 fluence of treatment.

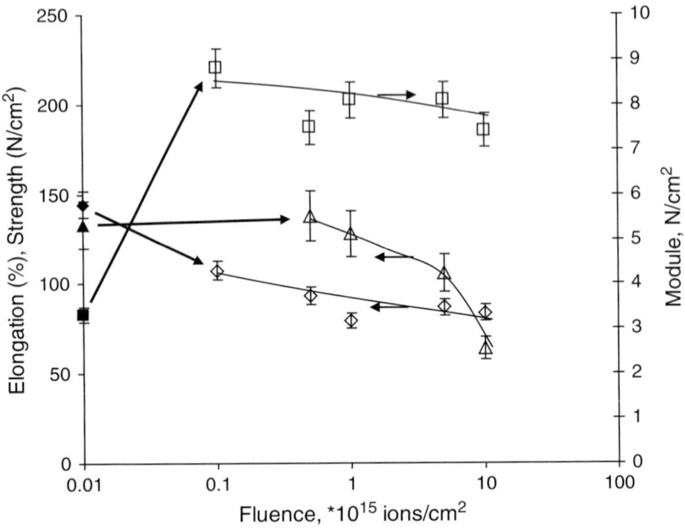

Fig. 4.53. Mechanical properties of polyethylene (LDPE) film (50 μm thickness) after PIII with 20 keV nitrogen ions: strength (rhomb), elongation at break (triangle), modulus (cubic). Full signs correspond to untreated LDPE.

etching rate and high crosslinking rate such as polyethylene and polystyrene. For polymers that have high etching and depolymerization rates, the mechanical properties do not change significantly (Fig. 4.54). It is observed for polymers such as PTFE and polyether.

At the end of analysis the structure of the ion beam implanted polymer can be drawn like a layer-to-layer system.

1st layer – thin first layer characterized by oxygen-containing groups (or other compounds as a result of reactions with active compounds of environment) bonded with destructed polymer macromolecules at low fluence or carbonized structure at high fluence. The structural changes in this layer are formed by direct action of penetrating ions and environmental conditions which begin to alter the polymer upon the first ion penetration.

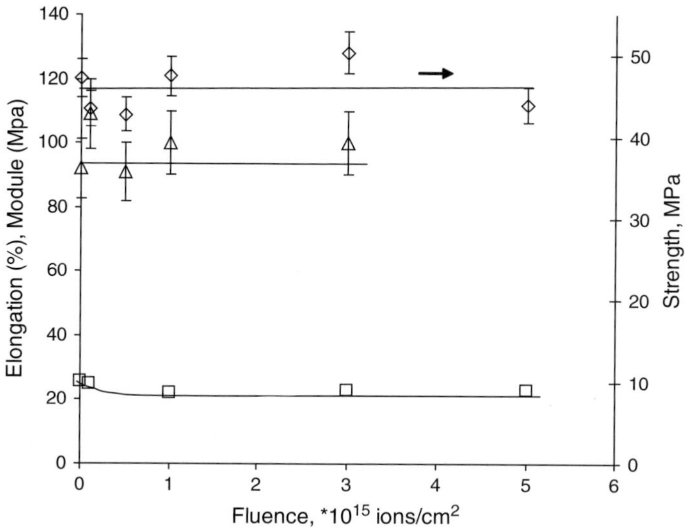

Fig. 4.54. Mechanical properties of PTFE film (20 mm thickness) after PIII with 20 keV nitrogen ions: strength (rhomb), elongation at break (triangle), modulus (cubic).

2nd layer – highly destructive structure, the thickness of this layer corresponds to ion penetration depth, the layer is highly carbonized. The structure of this layer is formed by spattering target atoms, high density of defects, high temperature and high pressure processes. This layer is attributed to the main result of structure transformation at ion beam implantation.

3rd layer – crosslinked layer or depolymerized layer (depends on the type of polymer). The structure of this layer is formed by free radical reactions, which can be described in terms of radiation chemical processes at low density of defects, low temperatures and absence of aggressive environment. The structure of macromolecules in such a layer does not change significantly.

4th layer – unmodified bulk polymer layer which is not changed.

An example of layer-to-layer structure for polyethylene is presented in Fig. 4.55. The structure changes from layer

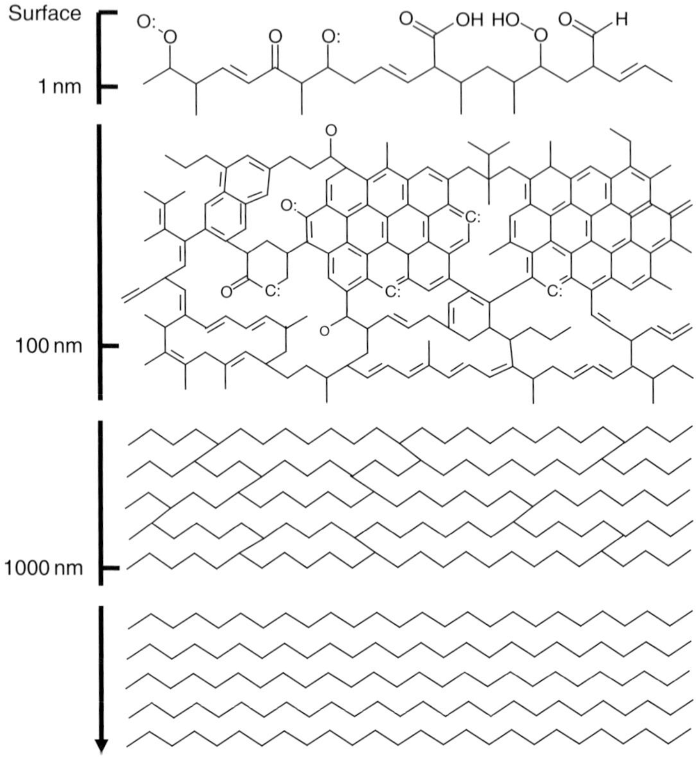

Fig. 4.55. Layer-to-layer structure of polyethylene after ion beam implantation. The example of structure is presented for ion beam implantation with 20 keV nitrogen ions.

to layer gradually. The sharp boundary is absent. In spite of this, the boundary between layers is rough, because the ion beam implantation process has stochastic character without uniform distribution of ions on the polymer surface. Areas with variable structure changes as well as the deeper and shallower modified layer can be found on the polymer surface.

This unique structure on the top of the polymer brings about unusual properties, which can be used for different applications.

References

1. D. Fink, *Fundamentals of Ion-irradiated Polymers*, Springer, Berlin, 2004, 404pp.
2. A.C. Ferrari, J. Robertson, Interpretation of Raman spectra of disordered and amorphous carbon, *Phys. Rev.* B61 (2000) 14095–14107.
3. F. Tuinstra, J.L. Koenig, Raman spectra of graphite, *J. Chem. Phys.* 53 (1970) 1126.
4. N.J. Harrick, *Internal Reflection Spectroscopy*, John Wiley & Sons Inc., 1967.
5. V.B. Odzhaev, I.P. Kozlov, V.N. Popok, D.V. Sviridov, *Ion Implantation of Polymers*, Belorussian State University, Minsk, 1998, 196pp.

5 Wetting

The interaction between liquid and polymer surface is described by the surface energy of polymer and liquid. One of the experimental methods of the interaction analysis is wetting of polymer by liquid drop. The form of liquid drop without any external interactions is spherical. In contact with the solid surface the form of drop deforms depending on surface tension. The value of deformation is characterized by the wetting angle Θ. The wetting angle is involved in an equation connecting γ_{LP} (the surface tension between liquid and polymer), γ_{LG} (the surface tension between liquid and gas environment over liquid drop) and γ_{GP} (the surface tension between gas environment and polymer):

$$\gamma_{GP} = \gamma_{LP} + \gamma_{LG} \cdot \cos\theta \tag{5.1}$$

When the wetting angle is high (Fig. 5.1), the interaction between liquid and polymer is weak, drop has a form close to spherical, and the wettability of the polymer surface is poor. When the wetting angle is low, the interaction between liquid and polymer is strong, the drop spreads on the surface, and the wettability is good.

The polymer/liquid interaction is described by a wide number of different kinds of intermolecular interactions between molecules of polymers and liquids. More detailed analysis of intermolecular interaction can be found in [1, 2]. In wettability, the measurement all types of intermolecular interaction is not possible. Usually, the complex interaction is simplified to two kinds of interactions, which are considered in different theoretical models of wettability. We use the Owens–Wendt–Rabel–Kaelble model where the intermolecular interaction is split into two components: polar and dispersic interactions. This model describes well the

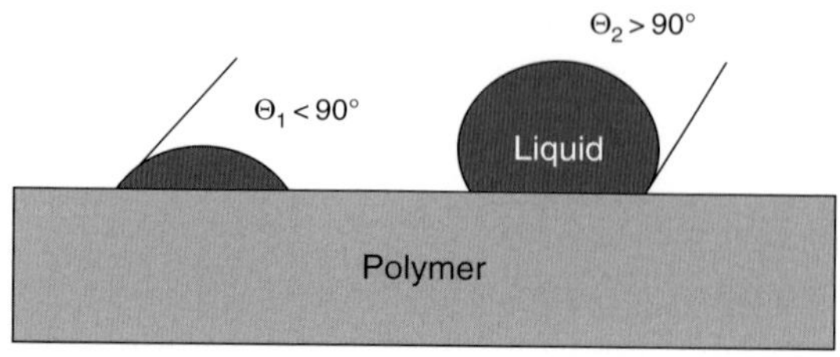

Fig. 5.1. Drop of liquid on polymer surface with low and high wetting angle.

structure transformations in polymers under ion beam and it is useful for an analysis of interactions with organic substances. In this model, the surface energy of the polymer is expressed as:

$$\sigma_s = \sigma_s^p + \sigma_s^d \tag{5.2}$$

and the surface energy of liquid is expressed as:

$$\sigma_l = \sigma_l^p + \sigma_l^d \tag{5.3}$$

Based on this splitting, the wettability measurements with two or more liquids could be used for calculating polar and dispersic parts of polymer surface energy. The calculation requires a linear regression solution of the equation for σ_s^p and σ_s^d:

$$\frac{(1+\cos\theta)\cdot\sigma_l}{2\cdot\sqrt{\sigma_l^p}} = \sqrt{\sigma_s^p}\cdot\sqrt{\frac{\sigma_l^p}{\sigma_l^d}} + \sqrt{\sigma_s^d} \tag{5.4}$$

The calculations need known parameters of liquid energy: σ_l, σ_l^p and σ_l^d. Table 5.1 contains such parameters for some liquids, which are used for wetting angle measurements. When choosing a liquid it is necessary to know that the liquid does not react with the polymer and the polymer does not swell in the liquid. In the case of chemical reaction or diffusion of liquid into polymer, the wetting data could be incorrect [3].

Table 5.1. Full surface energy and its parts for some liquids which are suitable for wetting angle measurements of polymers

Liquid	σ_l, MJ/m^2	σ_l^d, MJ/m^2	σ_l^p, MJ/m^2
Water	72.3	18.7	53.6
Glycerol	65.2	28.3	36.9
Formamide	59	39.4	19.6
Ethylene glycol	47.5	29.3	18.2
Diiodo methane	50.8	48.5	2.3
Dimethylsiloxane	19.0	16.9	2.1
Tricresyl phosphate	40.9	39.2	1.7
α-Bromonaphthalene	44.6	44.0	0.6

The measurement of wetting angle could be performed with a microscope. The volume of the drop must be small enough to exclude the influence of gravity on its form. The microscope must be equipped with a mirror and scale for horizontal view and dimension measurement of the drop. The wetting angle can be calculated from drop dimensions (Fig. 5.2):

$$\tan\theta = \frac{\sqrt{2rh - h^2}}{r-h} \quad \text{for } \Theta[\tau\sigma] > 90^\circ, \tag{5.5}$$

or

$$\tan\theta = \frac{2rh}{r^2 - h^2} \quad \text{for } \Theta < 90^\circ, \tag{5.6}$$

There are commercial devices for determining wetting angle measurement. Usually, such a device has a video camera

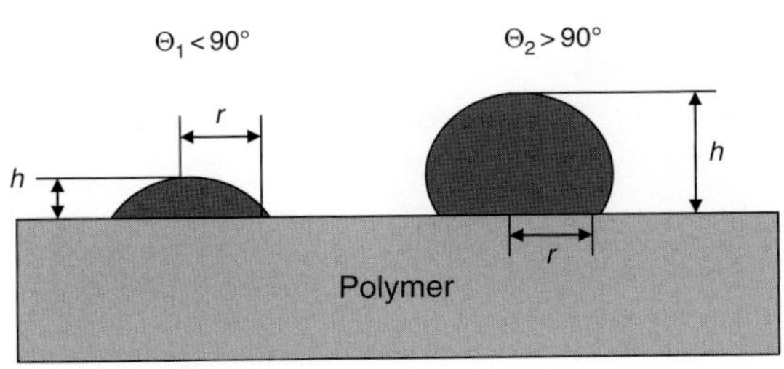

Fig. 5.2. Drop of liquid on polymer surface for calculation of wetting angle by dimensions of the drop.

and the drop form could be analyzed using special software. The wetting angle could be determined from analysis of the drop form.

Mainly, polymers have low surface energy when compared with the surface energy of metals. Usually, the polar part of surface energy is lower than the dispersic part. This means that a water drop has high wetting angle on an untreated polymer surface. For example, the wetting angle of a water drop on polyethylene (Fig. 5.3) is about 90° and on polytetrafluorethylene it is about 120°. An extremely high wetting angle of 150°–160° is observed on expanded polytetrafluorethylene (ePTFE), which has a fiber structure and water drop contact with polytetrafluorethylene fibers at some points and other water drop surface contacts only with air. Sometimes, the water drop could not be placed on the ePTFE surface using a syringe needle.

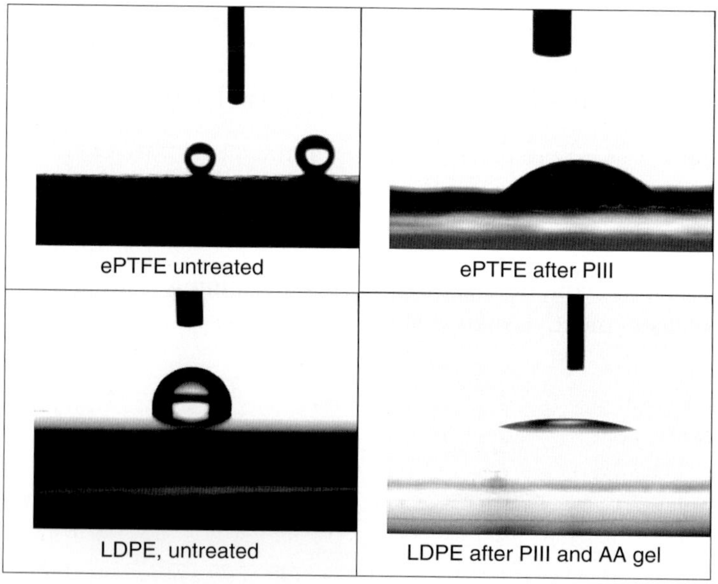

Fig. 5.3. Photographs of water drops on polyethylene and expanded polytetrafluorethylene (ePTFE) surfaces before and after PIII with post-treatment by acrylamide gel.

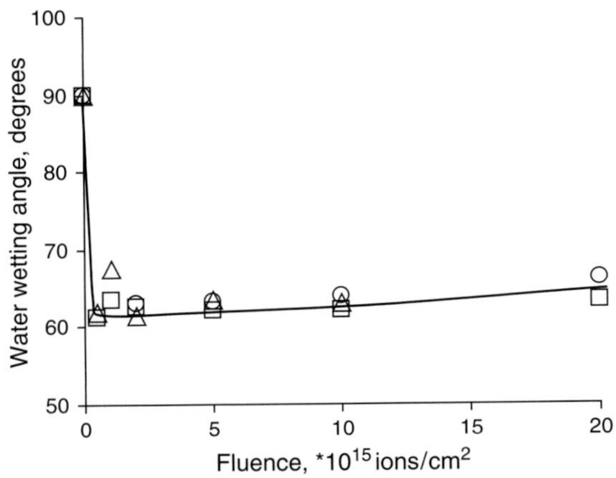

Fig. 5.4. Wetting angle of water drop on polystyrene surface in dependence of ion beam implantation dose for 20 keV (circle), 10 keV (cubic) and 5 keV (triangle) energy of nitrogen ions. The experiment was done in cooperation with K. Mizuno and B.K. Gan.

After ion beam implantation the wetting angle decreases significantly. It was observed for polyethylene [4–6], polystyrene [7, 8], polytetrafluorethylene [9], silicon rubber [10], polyurethane [11], polyamide [12], polycarbonate and polyimide [13]. For example, the wetting angle of a water drop on a polystyrene surface decreases from 90° to 60°–62° after ion beam implantation (Fig. 5.4). The fluence dependence of the wetting angle is not critical. Also the energy of ions is not critical to achieve such low wetting angles. The water wetting angle has a slight tendency to increase for high fluence treated polystyrene due to high carbonization of the surface layer.

Based on wettability measurement with four liquids on polyethylene, the calculation of surface energy and its components shows an increase to full energy (Fig. 5.5). The polar part of surface energy increases more than the dispersic part. This means that the polyethylene surface after ion beam treatment becomes more polar, and polar liquids wet polyethylene better than before treatment.

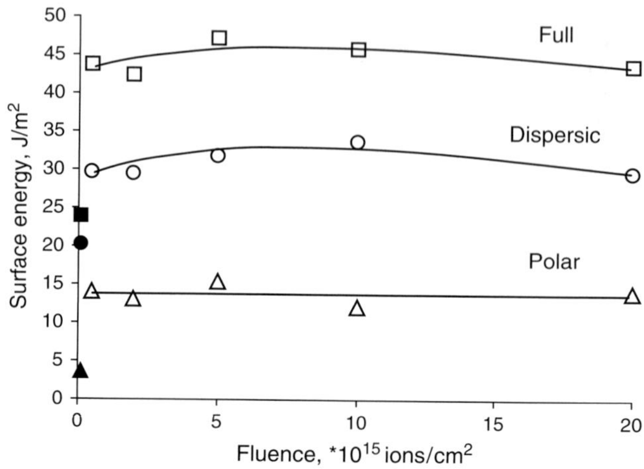

Fig. 5.5. Surface energy of polyethylene (LDPE) film after nitrogen PIII of 20 keV, measured after 2 weeks from PIII treatment.

The wettability of the treated polymer surface is unstable. After treatment the wettability changes are maximal. With ageing the wetting angles increase by asymptotical law, but the wetting angles of the untreated surface are not achieved (Fig. 5.6). If the water wetting angle is about 20°–30° just after implantation, the stable water wetting angle reached after one week is about 60°–70°. Such an angle is observed for different kinds of polymers after ion beam implantation and after stabilization of the wettability drift. The surface energy decreases also with ageing (Fig. 5.7). The polar part of energy decreases more quickly than the dispersic part. The surface energy and its parts change similarly for different polymers and reach approximately the same values after ageing.

The reason why wettability changes with time is connected with the instability of the chemical structure of the polymer surface after implantation. A high concentration of free radicals and instable chemical groups with high polarity cause a high increase of the surface energy and especially polarity of the polymer surface. With time the chemical reactions transform the unstable structure of the polymer

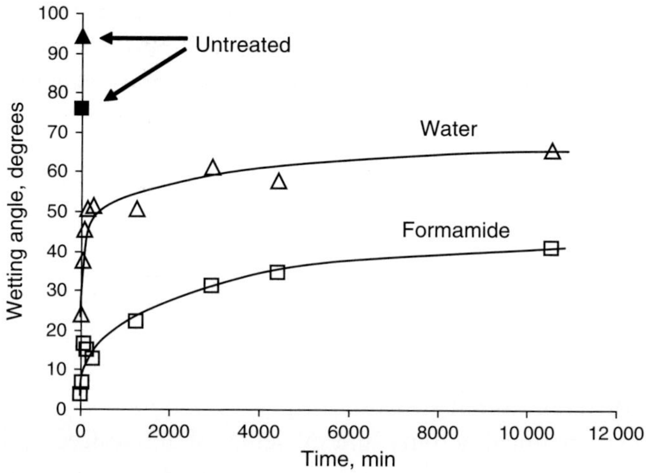

Fig. 5.6. Water and formamide wetting angles of polyethylene surface after PIII (fluence is 10^{16} ions/cm^2) with time of ageing.

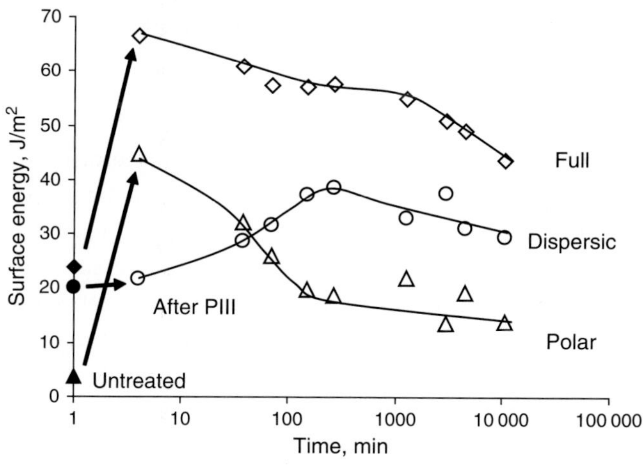

Fig. 5.7. Surface energies of polyethylene after PIII (fluence is 10^{16} ions/cm^2) with time of ageing.

surface into a stable structure which is characterized by a low concentration of free radicals and a high concentration of oxygen-containing groups which are stabilized by the presence of crosslinked carbonized structures with low mobility of macromolecules. A similar wettability effect is observed at plasma modification of the polymer surface, when wettability is significantly improved after treatment. But in the case of plasma treatment, the polymer surface layer does not contain a carbonized structure and polymer macromolecules can move into bulk layers of polymer that decrease the wettability up to initial untreated surface values.

The structure of the surface layer after ion beam implantation is similar for different polymers, therefore the wetting angles and the surface energy are the same independent of the different initial structure of polymers.

The chemical activity and presence of active free radical groups on a freshly treated surface can be used to obtain a stable, good, wet surface of the polymer. The polymer surface after ion beam implantation could be activated by active hydrophilic compounds such as acrylates or sugars. For example, the water wetting angle for untreated polytetrafluorethylene is equal to 110°–120°. If the polytetrafluorethylene surface is treated with PIII, the water wetting angle could reach 60°–80° after stabilization (Fig. 5.8). The water wetting angle on plasma treated polytetrafluorethylene reverts to 100° after stabilization. If the same polytetrafluorethylene surface after PIII is treated with acrylamide or sugar solution, the water wetting angle decreases by up to 30°. The reason for such a low angle is the formation of acrylamide gel on the polytetrafluorethylene surface. The acrylamide monomer can react with free radicals of the activated polytetrafluorethylene surface to form a crosslinked hydrogel layer on the top. This layer is hydrophilic and stable. A similar hydrophilic layer is formed by sugar, which can react with free radicals of polytetrafluorethylene. Such structures are stable and the wetting angle is stabilized for a long time. The same results were observed for different kinds of polymers such as polyethylene, expanded polytetrafluorethylene, polypropylene, polystyrene, and different kinds of polyurethanes.

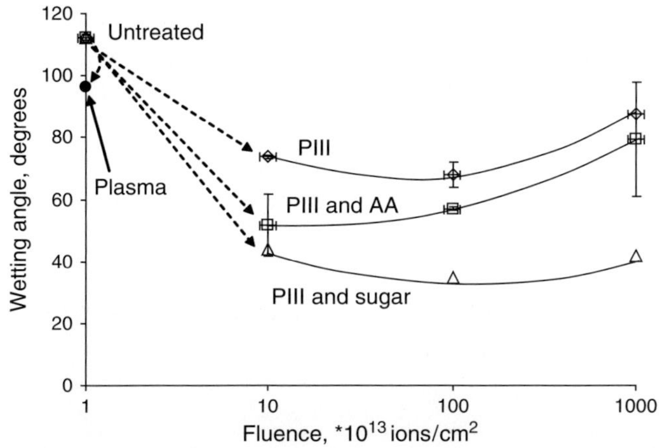

Fig. 5.8. Water wetting angles of PTFE after plasma (no bias) and after PIII with post-treatment by acrylamide (AA) and sugar.

Wettability is an important characteristic for polymer materials and in some applications a good wet surface is necessary. Ion beam implantation is used for the regulation of wettability of polymer materials for different applications. Let us consider some examples.

Dish washing machines have baskets for dishes, glasses, spoons, forks and knifes. Due to green technology the baskets are made from polymers, for example polypropylene. If the basket surface were hydrophobic, water drops would adhere to the surfaces of dishes, glasses, spoons, forks and knives and after drying these items would contain residual salt spots. Such spots are unwanted and modifications would be needed to counter this effect.

The polypropylene baskets are treated with PIII and afterwards with acrylamide and its analogue N-(hydroxymethyl)-acrylamide, which has a less toxic effect. After treatment the baskets are washed with water to completely remove residual monomers of acrylamide. The wetting angle using water after PIII is 65° and after PIII with acrylamide gel is 34° (Fig. 5.9). Such an angle is sufficient for good wetting with water and to remove water drops from the hydrophilic surfaces of the dishwasher's contents.

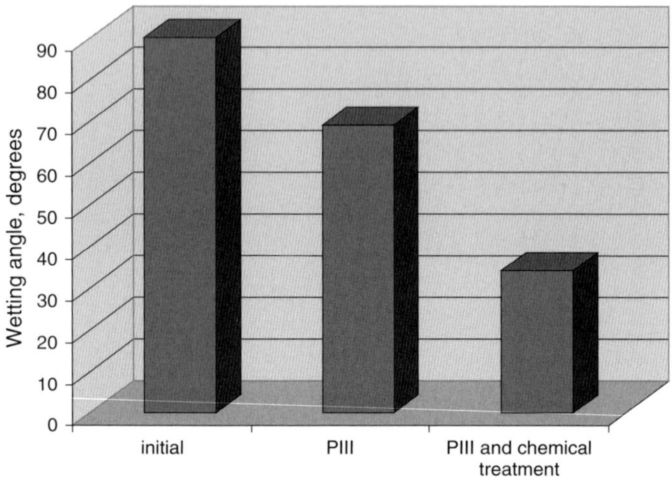

Fig. 5.9. Wettability of washing machine polypropylene basket after PIII and acrylamide post-treatment.

Wettability is responsible for friction if two surfaces are moving in a liquid environment. An example of the influence of wettability is polytetrafluorethylene in water. If a metal surface moves on dry polytetrafluorethylene, the friction is low. If the same metal surface moves on polytetrafluorethylene in water, the friction is high. In some cases, the friction on a polymer surface in water solution can cause problems. For example, a urine catheter used for opening a urine channel must be slippery enough to exclude the sensation of pain. Modern catheters are made with a polymer coating. To provide low friction, lubricants based on polyether and salts of fatty acids are used. At friction the lubricants are released from the catheter coating resulting in low friction. But in some cases the application of lubricants is not possible due to the sensitivity of organisms.

We modified the polyurethane coating of a catheter using PIII. After treatment the polyurethane surface was treated with acrylic monomers to synthesize acrylic gel on the surface. The wetting angle of the treated sample with the gel layer was about 30°. The friction coefficient of polyurethane in water solution decreases from 0.8 to 0.2.

Thin polymer film has a wide number of applications such as painting, coating, printing and so on. For such applications the instability of thin liquid film on substrate is one of the critical factors in technological processes. In connection with the importance of technological application the theoretical aspects of the instability of the thin liquid film have fundamental significance. A behavior of a thin liquid layer on the substrate can be described in terms of the classical mechanics of Newton liquid. The fluid film on the substrate can be considered as a large drop, which is squeezed to its own thickness. Following this approach, the height e_c of the drop flatted by gravity equals [14, 15]:

$$e_c = 2\sqrt{\frac{\gamma}{\rho g}} \sin\left(\frac{\theta}{2}\right) \tag{5.7}$$

where γ is the fluid surface tension, ρ is the fluid density, g is the gravity constant, and θ is the contact angle of the fluid drop with a substrate. The drop with height e_c is in equilibrium and stable on the substrate. If the drop height is smaller than e_c, the drop becomes unstable. The drop returns to the equilibrium state with height e_c. In this case, the edge of the drop, where surface tension forces are stronger due to large surface curvature, moves to the center. This process is called "dewetting". The dewetting process can be easily observed, for example, when oil is placed on a frying pan coated with Teflon or when rain falls on a car window.

For example, the dewetting process is observed for polystyrene film on a polytetrafluorethylene surface (Fig. 5.10). The polystyrene film dewets on the polytetrafluorethylene surface and forms separated drops. Size and distribution of drops depend on thickness of the polystyrene film. Such a polytetrafluorethylene surface cannot be used for printing processes due to low wettability by organic substrates. After PIII the surface energy of the polytetrafluorethylene is increased and close to the surface energy of hydrocarbon polymers like polystyrene (Fig. 5.11). In this case, the polar part of the surface energy is less important than the dispersic part, which is responsible for sufficient

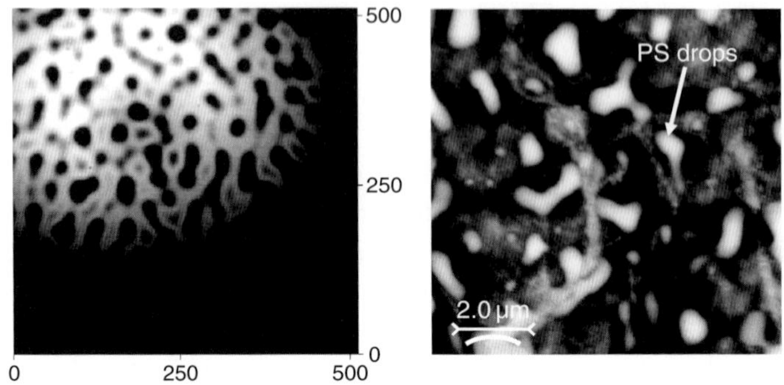

Fig. 5.10. Optical microphoto and AFM image of PTFE surface with dewetted polystyrene film of different thickness. Images were taken by P. Volodin.

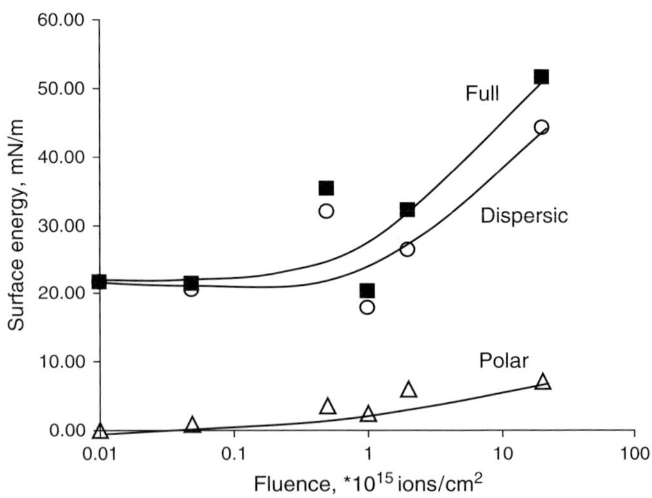

Fig. 5.11. Surface energy and its polar and dispersic parts of PTFE film after PIII with 20 keV energy nitrogen ions. The experiment was done in cooperation with B.K. Gan.

interactions with hydrocarbon polymers. After ion beam implantation, the polystyrene film is stable on modified polytetrafluorethylene and the effect of dewetting is not observed.

References

1. I.G. Kaplan, *Intermolecular Interactions: Physical Picture, Computational Methods and Model Potentials*, 380 pp., Wiley, 2006.
2. A.T. Kozulin, Yu.S. Klyachkin, A.V. *Kondyurin, Vibrational Spectra and Intermolecular Interactions*, Ekaterinburg, 1999.
3. M. Manso, A. Valsesia, M. Lejeune, D. Gilliland, G. Ceccone, F. Rossi, Tailoring surface properties of biomedical polymers by implantation of Ar and He ions, *Acta Biomaterialia* 1 (2005), 431–440.
4. J.S. Chen, Z. Sun, P.S. Guo, Z.B. Zhang, D.Z. Zhu, H.J. Xu, Effect of ion implantation on surface energy of ultrahigh molecular weight polyethylene, *J. Appl. Phys.* 93(9) (2003), 5103–5108.
5. K. Rockova-Hlavackova, V. Svorcik, L. Bacakova, B. Dvorankova, J. Heitz, V. Hnatowicz, Bio-compatibility of ion beam-modified and RGD-grafted polyethylene, *Nucl. Instr. Methods Phys. Res.* B 225 (2004), 275–282.
6. A. Kondyurin, V. Karmanov, R. Guenzel, Plasma immersion ion implantation of polyethylene, *Vacuum* 64 (2002), 105–111.
7. A. Lacoste, J. Pelletier, Processing considerations with plasma-based ion implantation of polymers: theoretical aspects, limitations, and experimental results, *Nucl. Instr. Methods Phys. Res.* B 208 (2003), 260–266.
8. A. Kondyurin, B.K. Gan, M.M.M. Bilek, K. Mizuno, D.R. McKenzie, Etching and structural changes of polystyrene films during plasma immersion ion implantation from argon plasma, *Nucl. Instr. Methods Phys. Res.* B 251 (2006), 413–418.
9. G. Mesyats, Yu. Klyachkin, N. Gavrilov, A. Kondyurin, Adhesion of polytetrafluorethylene modified by an ion beam, *Vacuum* 52 (1999), 285–289.
10. Y. Suzuki, C. Swapp, M. Kusakabe, Ageing effects on wettability and structure of ion implanted silicone, *Nucl. Instr. Methods Phys. Res.* 846 (1990), 354–357.
11. L. Dejun, Z. Jie, G. Hanqing, L. Mozhu, D. Fuqing, Z. Qiqing, Surface modification of medical polyurethane by silicon ion bombardment, *Nucl. Instr. Methods Phys. Res.* B 82 (1993), 57–62.

12. R.K.Y. Fu, I.T.L. Cheung, Y.F. Mei, C.H. Shek, G.G. Siu, P.K. Chu, W.M. Yang, Y.X. Leng, Y.X. Huang, X.B. Tian, S.Q. Yang, Surface modification of polymeric materials by plasma immersion ion implantation, *Nucl. Instr. Methods Phys. Res.* B 237 (2005), 417–421.
13. N.V. Gavrilov, V.N. Mizgulin, R. Stinnett, A.V. Kondyurin, Modification of polymer films of PE, PTFE, PC, PI by pulse ion beams, *Khimicheskaya fizika i mesoskopiya* 1(1) (1999), 39–47.
14. R. Fondecave, F. Brochard-Wyart, Polymers as dewetting agents, *Macromolecules* 31 (1998), 9305–9315.
15. C. Redon, F. Brochard-Wyart, F. Rondelez, Dynamics of dewetting, *Phys. Rev. Lett.* 66 (1991), 715–718.

6 Adhesion

The problem of adhesion is well known for polymer materials. Every new polymer material and its device needs an estimation of adhesion ability and a selection of the surface modification method, for example ion beam treatment, for adhesion improvement in most cases. This problem usually grabs the attention of scientists and technologists. Commonly, there are about 20 theories of adhesion, which are based on different mechanisms of interface interactions. These are: mechanical theory; theory of wettability; theory of surface energy; thermodynamical theory; adsorption theory; theory of chemical bonds; electrical theory; electrostatic theory; diffusion theory; mechanical theory; theory of weak layer; and others [1–9]. Each theory describes the behavior of joint strength for different adhesion phenomena, adhesion joint load and test, ageing and aggressive media. Therefore, the selection of modification technique for a new adhesion joint must be made on the basis of consideration of all adhesion theories, which need great effort or wide experience. How can the ion beam technique be estimated as the best method of adhesion improvement for real polymer material?

Our view on the problem of adhesion is based on structure inhomogeneity of the adhesion joint in the interface region. The structure means the multilevel structure organization of polymer. Every polymer material has a sufficiently complex structure, which is true even for simple polymers with few additives. Some structure levels can be found in polymer material. The structure classification is based on different interactions and roles in material properties. Usually, one can find the four following levels (Fig. 6.1): molecular level (molecular groups, intermolecular interactions), topological level (macrochain branches, molecular weight,

Fig. 6.1. Multilevel structure of polymer materials: molecular, topological, crystalline and macrostructure.

crosslinks), super-molecular level (crystallinity, ordering, lammeles) and macrolevel (fillers, particles, gas bubbles, surface roughness). Such structure division helps to understand the role of individual physical/chemical phenomena in properties of polymer composition. Depending on the kind and composition of polymer materials, the properties of the different polymer materials are determined by different structure levels. In terms of the structure levels, the adhesion joint is a break of the structure homogeneity. If we move from one material through an interface to another material, the structure has to jump the boundary between two jointed materials. If the structure level of the polymer materials is responsible for strength, the break of such a level on the interface region is responsible for strength of the adhesion joint. There are three variants of structure changes at the crossing of the interface region: structure changes smoothly, structure has a positive jump and structure has a negative jump (Fig. 6.2). Only in the last case

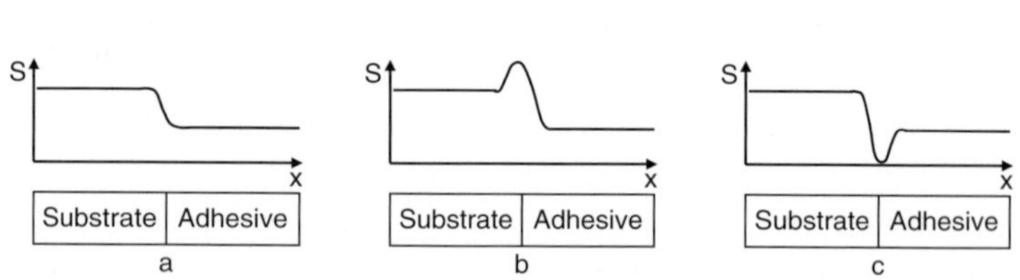

Fig. 6.2. Possible structure jumps in interface region. Cases (a) and (b) correspond to ideal adhesion joint. Case (c) corresponds to weak adhesion joint of low strength.

does the adhesion joint need improvement. Let us consider some different examples:

1. If the strength of jointed polymers is determined by relaxation processes of macromolecules caused by intermolecular interactions at mechanical load, then the interface region has to be filled with interlinked macromolecules of both polymers so strong that any mechanical load cannot move macromolecules away to different phases (Fig. 6.3a). The formation of the adhesion joint must provide an interpenetration of macromolecules through an interface with a sufficiently strong intermolecular interaction between macromolecules of different kinds of materials. This case can be described by thermodynamical theory, mechanical theory and diffusion theory.
2. If jointed polymers have a network of crosslinked macromolecules and contain low molecular components such as plasticizers, then the network is responsible for strength of the polymers (Fig. 6.3b). The crosslinks of the polymer network in the interface region must be generated at the formation of the adhesion joint. This case can be described by chemical theory of adhesion.
3. If jointed polymers are highly filled materials with inorganic hard particles, then particle distribution is an important factor for mechanical behavior of the materials (Fig. 6.3c). The disturbance of the particle distribution, caused, for example, by sedimentation processes or a speed distribution effect at molding of filled adhesive, forms the high concentration of particles and low concentration of polymer matrix in the adhesive near-interface layer. The strength of this layer is less than for bulk polymer and the strength of the adhesion joint is less than the maximum available.

In which case can the surface modification be useful for improvement of adhesion? Only in the second case is the surface modification able to generate special active groups for the formation of crosslinks in the interface region.

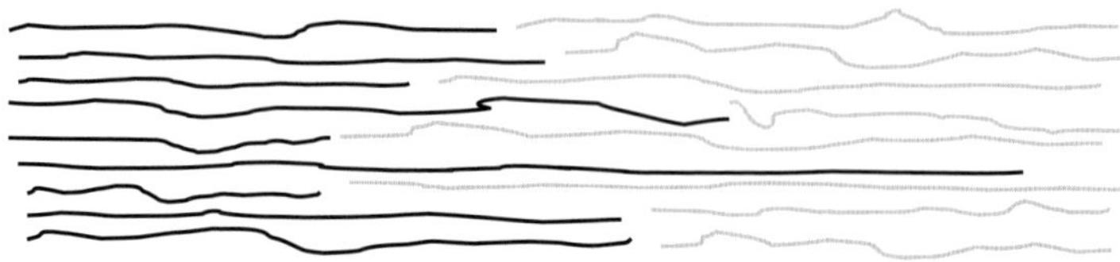

Fig. 6.3a. Interpenetrated macromolecules in interface region.

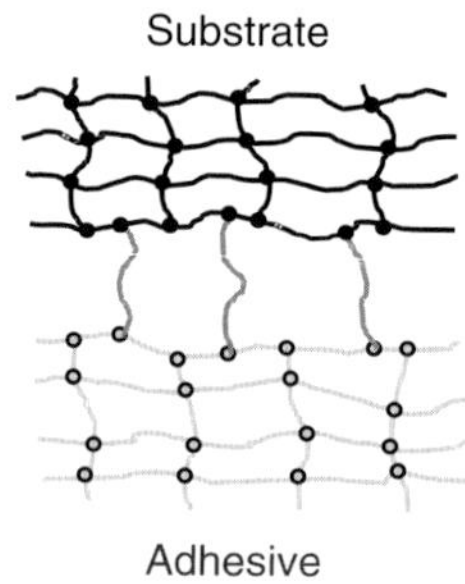

Fig. 6.3b. Crosslinked macromolecule networks in interface region.

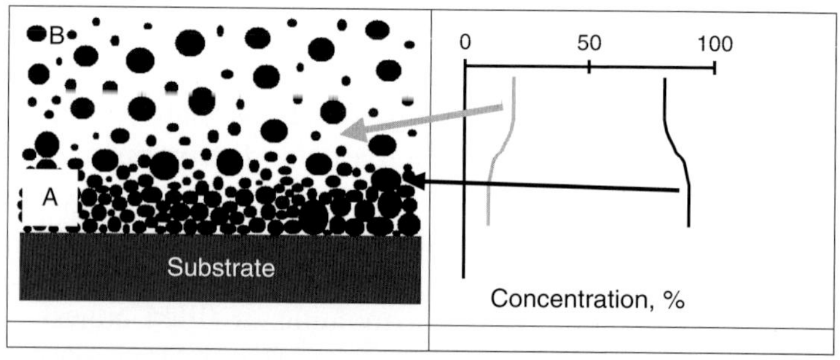

Fig. 6.3c. Filler inhomogeneity distribution in interface region. Gray is concentration profile of polymer matrix, black is concentration profile of filler.

In the other two cases, the modification of the polymer surface cannot improve adhesion and adhesion strength.

From these examples, one can see that the analysis of the adhesion joint must be done sufficiently precisely. "What structure level has jump?", "How strong is the structure jump?" and "How does this structure jump influence the mechanical properties of the polymer materials?" are the questions the answers to which show the reason for adhesion weakness (or strength) of the polymer joint. From this point of view, there is an ideal adhesion joint which has an homogeneous structure at the crossing of the interface, or at least a homogeneous structure on the main structure level. If there is no ideal adhesion joint and the break or jump of the important structure level is observed, we have to reconstruct the structure of materials to obtain a continuous structure at the crossing of the interface region. If this effect can be achieved by surface modification of one polymer material, only in this case does surface modification make sense and it can be selected for improvement of adhesion.

It should be noted that this is only one side of the adhesion problem. The other side is connected with stress distribution in the adhesion joint at mechanical load. In common, there are two cases of stresses: homogeneous and inhomogeneous distribution of mechanical stresses in the adhesion joint at test and operation. Both depend equally on conditions of the load and on differences of elastic properties of jointed materials. We don't consider here the stress analysis in the adhesion joints, but it must be taken into account at prognosis of adhesion joint behavior and its properties in real technological cases.

Let us consider some examples.

There is a well-known adhesion joint of polyethylene and its co-polymers with epoxy adhesives. One of the industrial applications of polyethylene co-polymers is EPDM rubber. This is a ter-polymer of ethylene, propylene and diene, which has perfect mechanical properties in the wide temperature region; this rubber is used in different fields from automobiles to rockets. But EPDM rubber doesn't adhere well to active epoxy or urethane adhesives.

EPDM rubber usually has a complex content: plasticizers, inorganic fillers, stabilizers, antioxidants and groups of vulcanizing agents. The rubber has polymer networks as well as epoxy and polyurethane adhesives. The network of rubber provides excellent mechanical properties sufficient for exploitation at different temperatures and in aggressive media. In the presence of a plasticizer of high concentration the polymer network is a requirement of high mechanical strength of the rubber. In the case of adhesion joint, the network is the most important parameter and crosslinks in the interface region between networks of rubber and adhesive must be formed.

The FTIR ATR spectra of EPDM rubber shows lines at the 1750–1710 cm^{-1} region, interpreted as vibrations of carbonyl groups. These groups are the result of the oxidation process of EPDM macromolecules in the surface layer under the action of oxygen in air, UV light and natural radiation. The effect of oxidation is observed for all kinds of polymer at storing. As a result of oxidation the carbonyl, carboxyl, aldehyde and hydroxyl groups are formed in the surface layer of rubber. These groups can react with epoxy groups of adhesives in the following ways:

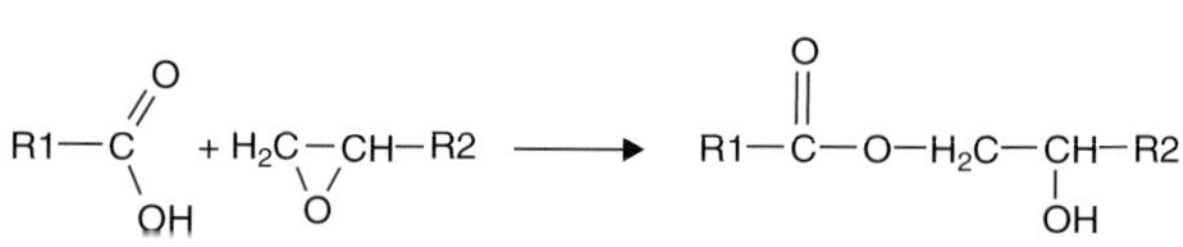
(6.1)

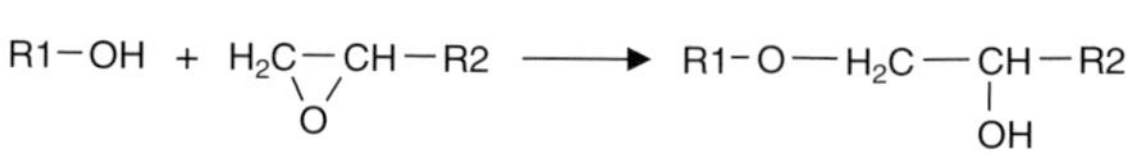
(6.2)

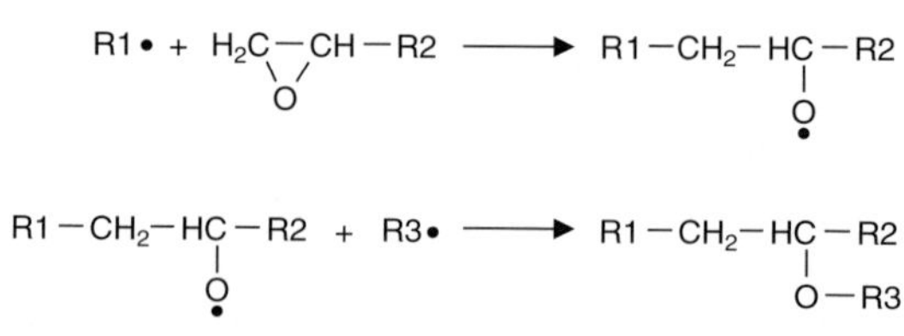
(6.3)

The result of the reaction can be observed by FTIR ATR spectra of the EPDM surface after exposition of the rubber in liquid epoxy resin without hardener (Fig. 6.4). After washing in toluene the EPDM rubber surface contains a small amount of crosslinked epoxy resin which is jointed

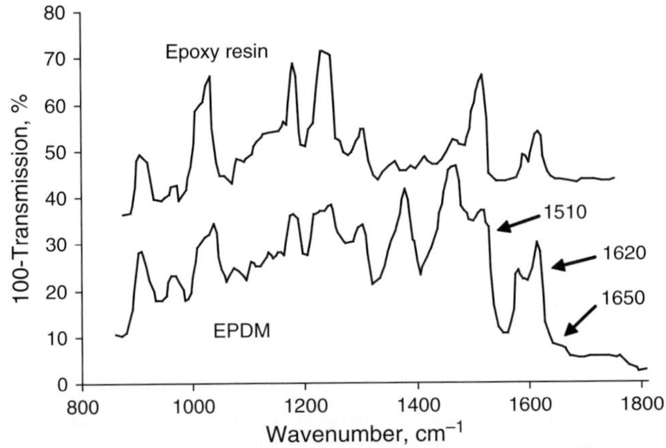

Fig. 6.4. FTIR ATR spectra of EPDM rubber after exposure in liquid epoxy resin without hardener. For comparison the spectra of liquid epoxy resin is shown on the top. Arrays show lines of the epoxy resin.

with the rubber network by one of the reaction in (6.1)–(6.3). The lines of 1620 and 1510 cm^{-1} are associated with vibrations of aromatic ring of epoxy resin macromolecule, and the line at 1650 cm^{-1} is associated with vibrations of the ester group as a result of the above reactions. Therefore, the critical structure level of this adhesion joint is the crosslinks between networks in the interface region, which are formed due to reactions between the epoxy group of adhesives and the oxygen-containing groups of the oxidized surface layer of the rubber. But is the concentration of the oxygen-containing groups enough for an ideal adhesion joint?

The answer can be found on the basis of the failure character of the adhesion joint. If the adhesion joint can be peeled on the interface boundary, the interface region is not strong enough and adhesion can be improved. If the adhesion joint is broken on one of the jointed polymers, the adhesion strength is maximal. But the failure character depends on mechanical properties of the jointed polymers, geometry of the adhesion joint and load at adhesion test. In some cases, the failure character can be attributed to a weak region of the adhesion joint. For our experiments both polymers have similar elastic and strength properties when we used elastic resin PDI based on polyisoprene-polydivinyl with end epoxy groups and the hardener SKN based on butadiene acrylonitryl resin with carboxylic end groups. The geometry of adhesion joint and load conditions gave us the possibility to use the failure character for the determination of the weakest region of the adhesion joint. In our case it was peeling test at low rate of load.

The analysis of the failure character is connected with the main structure level. How is it possible to determine the adhesion or cohesion character of failure? In any adhesion joint the interface boundary is not clearly determined. For example, a number of macromolecules can penetrate through the interface and be stored in the surface layer of the second polymer even at very low adhesion. Or in the case of high filler polymer the adhesion failure can go some distance from the initial interface, but the failure has adhesive character because the surface layer of adhesive can have less concentration of filler and lower strength. In both cases the failure has adhesion character because it is caused by the interface, and the failure surface layers contain the results of interface phenomena. The main problem at analysis of adhesion joint is to find the main structure level which is responsible for adhesion strength. In most cases, the structure level determines mechanical strength of material, and the same structure level can be responsible for adhesion strength and failure character.

In the case of EPDM rubber, the adhesion joint failure has adhesion character. It was supported by the solvate method and FTIR ATR spectra of failure surfaces. For the solvate method, the adhesion joint is swelled in binary solvent, in which both rubber and cured resin are swelled at the same volume. It was found for an excluding of tangential stresses in interface region. After swelling the adhesion joint on the interface boundary was very easily peeled off. The FTIR ATR spectra of failure surfaces (Fig. 6.5) showed that lines of EPDM rubber are observed only in the first side of failure and the lines of adhesives are observed

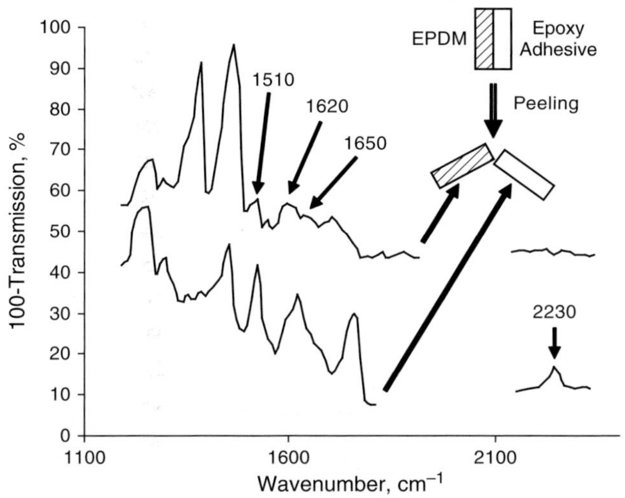

Fig. 6.5. FTIR ATR of failure surfaces of adhesion joint from EPDM rubber side and epoxy adhesive side.

only in the second side of failure. But very weak lines of epoxy resin can be recognized in spectra of the EPDM side. This means that failure occurs on the interface boundary; the network in the interface boundary is not sufficient for good adhesion despite some epoxy macromolecules reacting with EPDM oxygen-containing groups. In this case, an increase in oxygen-containing active group concentration on the EPDM surface can increase the adhesion.

The FTIR ATR spectra of EPDM rubber after ion beam treatment show an increase in carbonyl line intensity with fluence of ions (Fig. 6.6). The peak intensity in the 1750–1600 cm^{-1} region of spectra increases with fluence of ion beam treatment. The same increase in carbonyl line intensity is observed after both plasma discharge treatment and UV light treatment. The increase in carbonyl group concentration leads to an increase in chemical activity of the EPDM surface. After surface modification the spectral line intensity of the epoxy resin macromolecules jointed to the EPDM surface increases (Fig. 6.7). This means that the concentration of the epoxy resin macromolecules increases

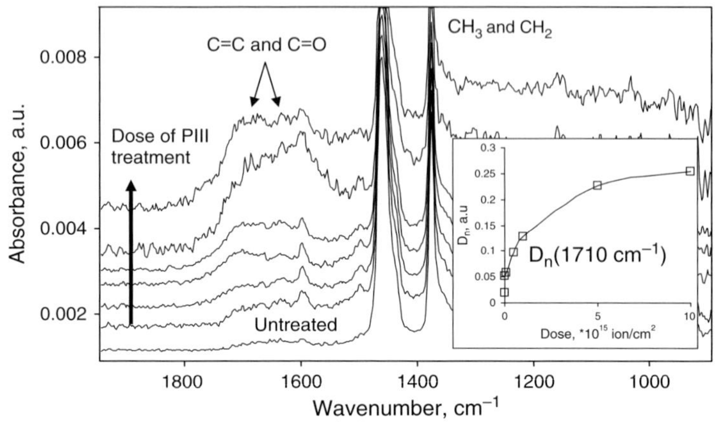

Fig. 6.6. FTIR ATR spectra of EPDM surface after ion beam modification. The fluence of ion beam treatment increases from bottom to top.

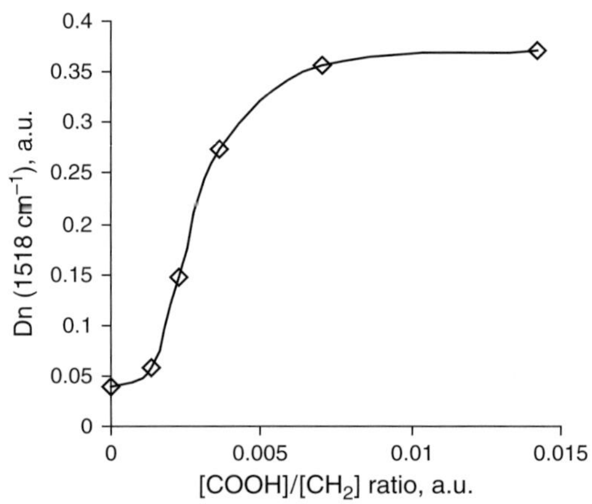

Fig. 6.7. Epoxy macromolecule line intensity in FTIR ATR spectra of EPDM with different concentrations of carbonyl groups in the surface layer.

and a denser network forms in the interface region of the adhesion joint. The strength of the adhesion joint increases with fluence of treatment corresponding to the increase of the network density in the interface region (Fig. 6.8).

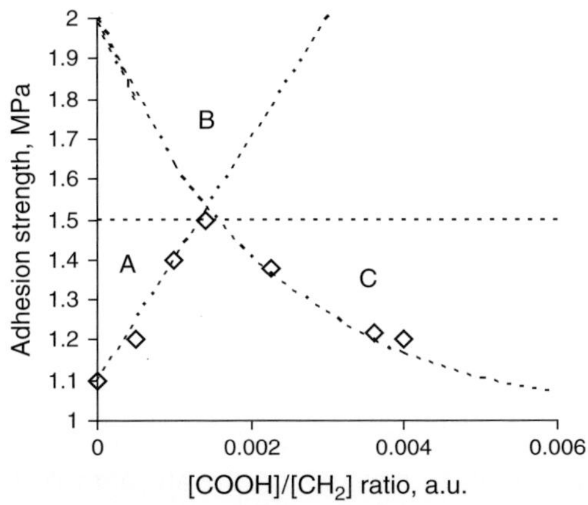

Fig. 6.8. Strength of EPDM adhesion joint depending on dose of surface modification of polymer substrate. Dashed curves are theoretical, rhomb signs are experimental.

Together with the effect of modification under different kinds of irradiation, there is the effect of surface layer degradation in the irradiated polymer and the strength of this layer can decrease with fluence of treatment. Related to fluence increase, the strength of the adhesion joint can decrease at high fluences. Such dependence of adhesion strength is common for any method of surface modification of polymer materials. The strength dependence can be expressed by part-broken function as [10]:

$$\sigma = \begin{cases} \sigma_1 + k_1 \cdot D & D < D_1 \\ \sigma_1 & D_1 < D < D_2 \\ \sigma_2 + \sigma_{02} \cdot e^{-k_2 \cdot D} & D > D_2 \end{cases} \tag{6.3}$$

where σ is the strength of the adhesion joint, σ_0 is the initial strength of the adhesion joint, σ_1 is the maximal strength of the adhesion joint which corresponds to strength of one of the jointed materials, k_1, k_2 and σ_{02} are parameters and D is the degree of modification.

Based on this adhesion strength model, some critical cases can be considered (Fig. 6.9):

1. The initial strength of a joint is close to the strength of one of the polymers ($\sigma_0 = \sigma_1$). There is an ideal adhesion joint without modification. Any surface modification can only reduce the adhesion strength. There is no need for surface modification.
2. The adhesive strength is close to the substrate strength ($\sigma_1 = \sigma_2 + \sigma_{20}$). Modification of the surface can increase the adhesion strength, but in some cases the destruction effect (k_2 is high) can reduce the strength of the surface layer and the ideal adhesion joint cannot be achieved.
3. Modification of the substrate surface does not increase the adhesion ability because the main structure level does not change at such kinds of modification ($k_1 = 0$). Modification can only decrease the adhesion strength.

Such a model of the adhesion joints was applied for adhesion joints of different kinds of polymers at UV light, plasma discharge and ion beam modifications [11–12].

Let us consider modification of the simplest polymers polyethylene (LDPE) and polytetrafluorethylene (PTFE) which have poor adhesion qualities to reactive epoxy and isocyanate adhesives.

LDPE and PTFE have poor thermodynamic compatibility with epoxy and isocyanate polymers, therefore thermodynamic theory, interpenetration of macromolecules and adsorption theories cannot be applied. After ion beam treatment the surface energy of LDPE and PTFE changes which leads to a decrease in wetting angles of substrates by adhesives in the liquid state. The wetting angle of LDPE by epoxy resin decreases from 60.4° for initial LDPE to 37.3° for LDPE after nitrogen ion beam treatment with an energy of 20 keV and a fluence of 3×10^{14} ions/cm^2. The wetting angle of PTFE by epoxy resin decreases from 96° for initial PTFE to 76° for PTFE after nitrogen ion beam treatment with an energy of 20 keV and a fluence of 10^{15} ions/cm^2.

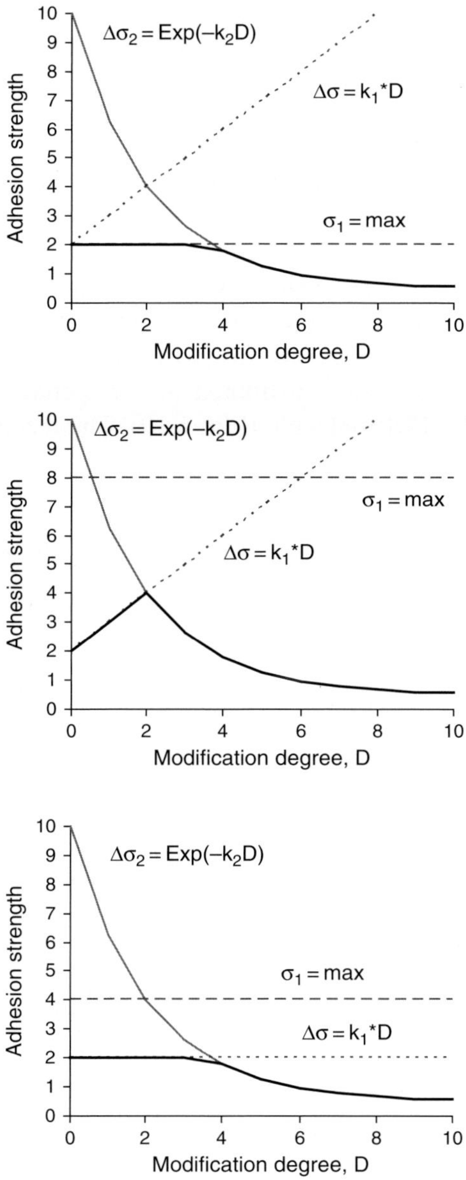

Fig. 6.9. Model of the adhesion strength in the dynamic of substrate surface modification. See text.

The appearance of the oxygen-containing groups after ion beam changes the chemical activity of the LDPE and PTFE surfaces – observed by experimenting with the exposition of treated LDPE and PTFE films in liquid toluene diisocyanate (TDI). After exposition of LDPE film in TDI the surface of LDPE contains the urea and urethane groups as a result of the chemical reaction of LDPE active groups with the isocyanate group of TDI. In FTIR ATR spectra it is seen by the appearance of amide lines at $\nu(C{=}O) = 1700–1750\,cm^{-1}$ and $\delta(N{-}H) = 1520–1550\,cm^{-1}$, the line of the TDI aromatic ring at $1610\,cm^{-1}$, and the line at $\nu(NCO) = 2285\,cm^{-1}$ attributed to the second isocyanate group in the TDI molecule (Fig. 6.10). The reactions occur as follows:

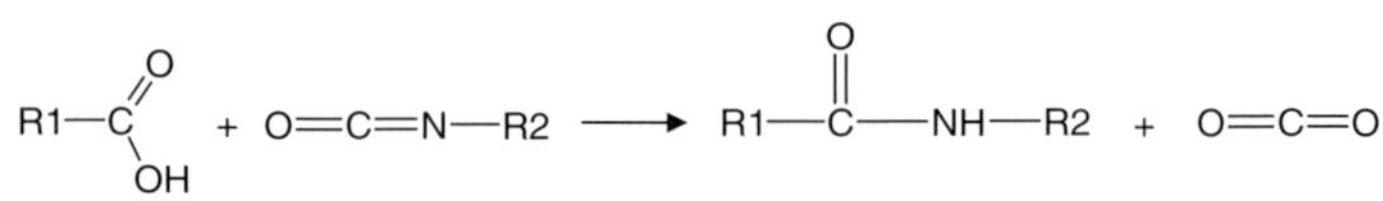

(6.4)

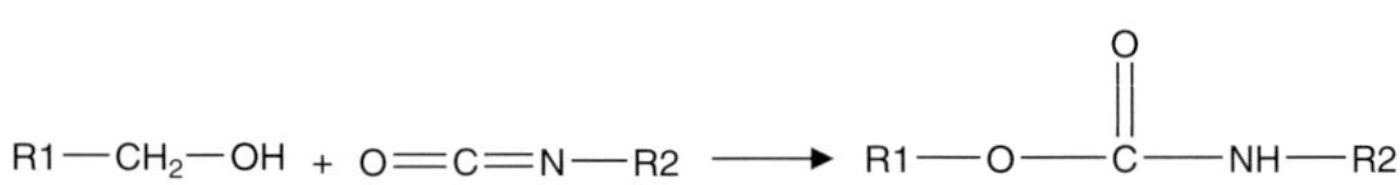

(6.5)

The intensity of such lines does not change after washing the LDPE surface with acetone and toluene. In the FTIR ATR spectra of the untreated side of the LDPE film such spectral lines are not observed as well as in spectra of untreated LDPE film after exposition in TDI. Therefore, modified macromolecules of polyethylene in the surface layer can react with the isocyanate group of TDI. Such reactions occur in the interface region between liquid adhesive and LDPE surface at the formation of the adhesion joint. After curing of isocyanate adhesive, the interface region contains crosslinks of polymer networks of the substrate and adhesive.

The same method can be used for analysis of the interaction of the modified LDPE surface layer and epoxy adhesive. After exposition of LDPE film in spectra of the

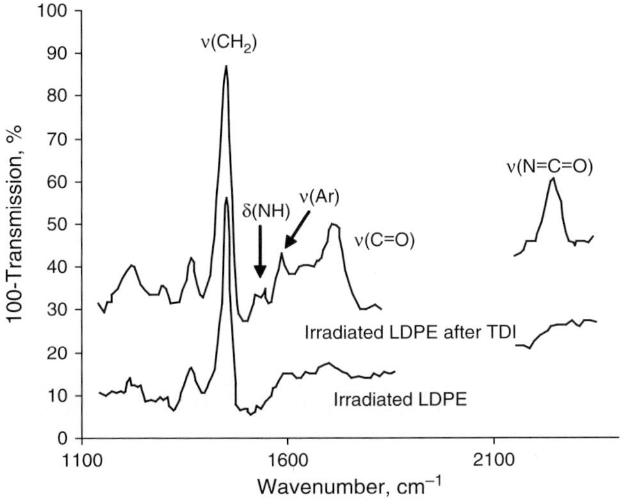

Fig. 6.10. FTIR ATR spectra of LDPE treated with ion beam and exposed to toluene diisocyanate (TDI).

surface layer the lines of 1250 and 1040 cm^{-1} interpreted as vibrations of epoxy macromolecules are observed on the modified side and unobserved on the untreated side. After washing the LDPE film in acetone and toluene the intensity of these lines does not change. This means that the epoxy resin reacts with oxidized macromolecules of LDPE and such reactions can occur in the interface region of the adhesion joint. The kinds of chemical reaction are similar to those presented for EPDM rubber in (6.1)–(6.3).

The surface of PTFE becomes active to isocyanate and epoxy compounds too (Fig. 6.11). After exposition of modified PTFE film in TDI the spectra of the surface show the appearance of the additional lines at $\nu(C{=}O) = 1720$ and 1700 cm^{-1} of the urethane group, $\nu(C{=}O) = 1670\ cm^{-1}$ of the urea group, 1610 cm^{-1} of aromatic ring of TDI, $\delta(NH) = 1563$, 1530–1520 cm^{-1} of the urea and urethane groups, $\delta(CH_3) = 1460$, 1430, 1390 cm^{-1} of TDI, and $\nu(NCO) = 2285\ cm^{-1}$ of the second isocyanate group of the TDI molecule. These lines remain after washing of the modified PTFE surface with solvents. The same effect is

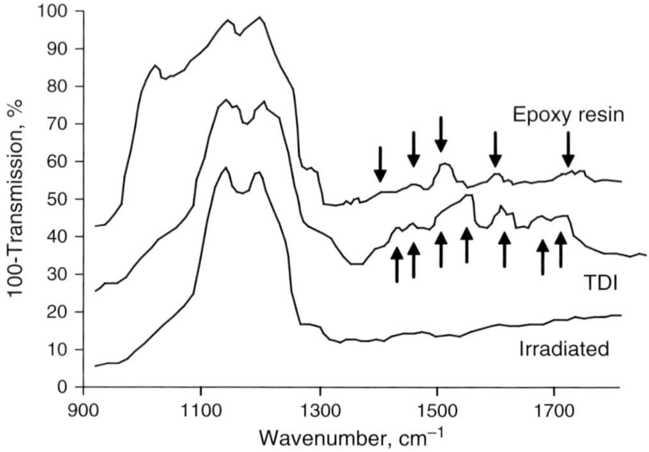

Fig. 6.11. FTIR ATR spectra of PTFE treated with ion beam and then exposed to TDI and epoxy resin.

observed at exposure of the modified PTFE film in liquid epoxy resin. Spectra of treated PTFE surface contains 920, 1468, 1480, 1520, 1610, 1630 cm^{-1} lines attributed to vibrations of epoxy resin macromolecules. These lines remain in the spectra of PTFE after washing with acetone and toluene. But if washing is done with sufficient mechanical effort, the modified PTFE layer can be removed together with attached epoxy resin. This effect is connected with weak interactions between macromolecules of PTFE, which can be easily removed or rearranged.

The strength of the adhesion joints of LDPE increases sharply with fluence of ion beam treatment (Fig. 6.12). The strength achieves a value, which corresponds to the strength of epoxy resin determined at the same geometry and test conditions. At next fluence increase the strength slightly decreases. At such changes of adhesion strength the failure character changes too. For initial LDPE the failure occurs on the interface boundary. The spectra of the failure surface show a clear division between the LDPE and adhesive surfaces. The failure of modified polymer joints is cohesive on epoxy adhesive which is observed even visually.

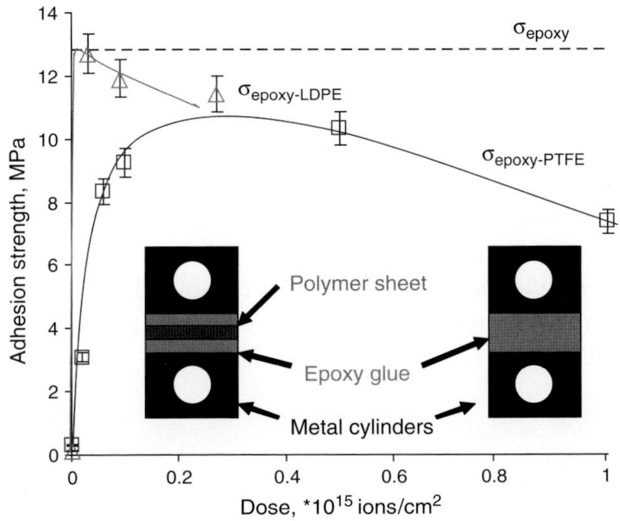

Fig. 6.12. Adhesion strength of LDPE (triangle) and PTFE (cubic) with the same epoxy adhesive (top line is adhesive joint strength without polymer sheet).

Adhesion ability of the modified LDPE does not change after washing with acetone and CCl_4. The maximal value of the adhesion strength is similar to and even higher than well-known methods of modification of polyolefin surfaces by chromic acid and corona discharge (8.3–9.0 MPa).

The same behavior is observed for the adhesion strength of PTFE. The adhesion strength increases with fluence of ion beam treatment. The failure character transforms from adhesive for untreated PTFE to cohesive for modified PTFE. The maximal strength of the adhesion joint (13.55 MPa) reaches the strength of the epoxy adhesive and this value is higher than the well-known adhesion strength of PTFE modified with potassium/naphthalene complex (11.0 MPa). The modified PTFE keeps its adhesion ability a long time after treatment. We observe the same adhesion ability after one year of storage of the modified PTFE under laboratory conditions before the adhesion joint formation.

The polymer networks' crosslinks of the epoxy adhesive with LDPE and PTFE in the interface region are stable because of chemical bonds. This means that the adhesion strength is stable in aggressive environments and temperature ranges which is available for jointed materials as for itself.

Therefore, after ion beam treatment the adhesion strength reaches its maximal value and the adhesion joint becomes ideal. A scientific problem has been solved. There is only one remaining question: what price of your polymer device is to use expensive ion beam implanter?

References

1. A.J. Kinloch, *Adhesion and Adhesives: Science and Technology*, Chapman and Hall, London, 1987.
2. N.A. de Bruyne, R. Houwink, *Adhesion and Adhesives*, Amsterdam: Elsevier, 1951.
3. D.D. Eley, *Adhesion*, London: Oxford University Press, 1961.
4. A.A. Berlin, B.E. Basin, *Fundamentals of Polymer Adhesion*, Khimiya, Moscow, 1974.
5. S.S. Voyutskii, V.L. Vakula, *J. Appl. Polym. Sci.* 7 (1963), 475.
6. D.R. Paul, S. Newman (Eds), *Polymer Blends*, Vol. 1, Academic Press, New York, 1978.
7. B.E. Basin, *Adhesive Strength*, Khimiya, Moscow, 1981.
8. J. Bjorksten, L.L. Yaeger, *Mod. Plast.* 29 (1952), 124.
9. V.L. Vakula, L.M. Pritikin, *Physical Chemistry of Polymer Adhesion*, Khimiya, Moscow, 1984.
10. A. Kondyurin, Yu. Klyachkin, Adhesion of UV-treated rubbers to epoxy adhesives, *J. Appl. Polym. Sci.* 62 (1996), 1–8.
11. G.A. Mesyats, Yu.S. Klyachkin, N.V. Garilov, V.N. Mizgulin, R.M. Yakushev, A.V. Kondyurin, Ion beam modification of polyethylene and adhesion to epoxy adhesive, *Vacuum* 47(9) (1996), 1085–1087.
12. G. Mesyats, Yu. Klyachkin, N. Gavrilov, A. Kondyurin, Adhesion of polytetrafluorethylene modified by an ion beam, *Vacuum* 52 (1999), 285–289.

7 Hardness

In some applications, a polymer is required to resist scratching. Examples include optical elements made from polycarbonate resins such as CR39 whose scratch resistance is measured with a test in which a sharp point is drawn over the surface and is subjected to an increasing normal force. The onset of damage is detected either through the measurement of the friction coefficient or by the measurement of acoustic emission. One of the methods of improvement in scratch resistance is an increase in the hardness of the surface layer by carbonization under ion beam implantation [1]. The carbonized surface layer has an indentation hardness higher than the hardness of the initial polymer and a different force-indentation curve. The thickness of the carbonized layer depends on the ion penetration depth. But the improved polymer surface layer could be thicker than the ion penetration depth due to the diffusion of radicals into the material beyond the end of the range of ions [2].

The increase in hardness was observed after ion beam implantation (by PIII) of nitrogen ions into polyethylene terephthalate (PET) sheet which is used for manufacturing plastic bottles [3]. Ion beam implantation has been used for increasing the hardness of bimorphic humidity sensors made from polyimide and polyethersulfone [4]. Ion beam implantation has been used to increase the hardness of high grade transparent optical lenses made from polycarbonate resin (CR39) [5, 6]. The ion beam treatment of polymers induces darkening which may compromise the lenses for some applications.

One of the important applications of polymers is for medical devices for use in the human body. Polymer materials have high biocompatibility, low toxicity and a range of mechanical properties that are similar to those of tissue materials. Surface hardness is an important mechanical characteristic of polymers [7–9] which is important in applications for the replacement of joints in the body. A number of biomedical applications require special improved scratch resistance hardness of the polymer surface. For example, ion beam implantation was used for hardness improvement of artificial hertz valve materials as PEEK [10]. A part made out of polymer usually operates at low mechanical load, far from breaking stresses; however, even a low mechanical load can cause a defect of the microstructure, which can progress at high loads into macro cracks and final destruction of the part. Prevention of the initial defect has such a strong influence on the exploitation life-time and thus in this case the medical device for which operation reliability is mandatory. An increase of surface layer hardness can prevent the structure defects which appear at low scratch load.

A low friction coefficient is important in some applications involving sliding contact. For example, the trobological properties of ultra-high molecular weight polyethylene (UHMWPE), HDPE and LDPE were improved by ion beam implantation. These materials can be used for total joint replacement prostheses (TJR) (e.g. acetabular cups and tibial components) [11–13].

Another example of an application requiring high scratch resistance is a stent delivery system as used for placing a stent in coronary or peripheral blood vessels [14]. A thin-walled polymer balloon is part of this stent delivery system. The balloon is folded, after which the metallic stent is crimped around the folded balloon to a very small diameter to enable the delivery of the system to very stenosed regions. Once the delivery system has been guided through the vascular system to its destination, increasing the pressure inside the balloon expands the balloon, opening up the stenosed blood vessel and enabling simultaneous deployment of the stent.

The balloon wall must be soft and thin to allow the medical device to be guided through tortuous vascular systems without damaging the vessel wall; the balloon wall must be strong enough to withstand the high inflation pressure and withstand potentially very high point forces caused by a possible high degree of calcification in hard lesions. It is therefore essential that the metal stent crimped on a folded balloon will not scratch the balloon thereby ensuring that the balloon will not fail under these circumstances. In order to satisfy these apparently contradictory requirements, a very thin surface layer of increased hardness was produced without significant change to the bulk properties of the balloon wall. Ion beam modification was found to be successful for increasing the hardness of the surface of Pebax [15] without affecting the interior of the thin wall of the balloon.

Infrared and Raman spectroscopies are useful when observing the changes in the structure of polymers after ion beam modification. For example, the structure changes in Pebax after ion beam implantation were observed by micro-Raman spectroscopy in which the laser exciting beam is confined to a thin surface layer of Pebax. The lines at 1645, 1446, 1381, 1305, 1121, 1074 cm^{-1} corresponding to the vibrations of polyamide/polyether Pebax macromolecules were observed in the spectra of the initial Pebax but disappear with increasing fluence of treatment (Fig. 7.1) and a wide peak in the 1300–1600 cm^{-1} region appears. The 1300–1600 region of the Raman spectrum contains information on the carbon-rich network that is produced by the modification. A peak at 1540–1580 cm^{-1} is named the G-peak and corresponds to carbon in the form of graphitic structures. The peak at 1330–1360 cm^{-1} is named the D-peak and the intensity of this peak increases with the degree of disorder of the graphitic structure. The high intensity of these lines in modified polymers can be explained by resonance effect in Raman spectra on graphitic structures coupled with basic breathing vibrations of aromatic rings.

At 20 keV ion energy the 1542 cm^{-1} peak corresponding to carbon vibrations in graphitic phase has a strong intensity. This means that a graphite structure appears as a result of the carbonization of the polymer surface layer. Following treatment with 30 keV energy ions at low

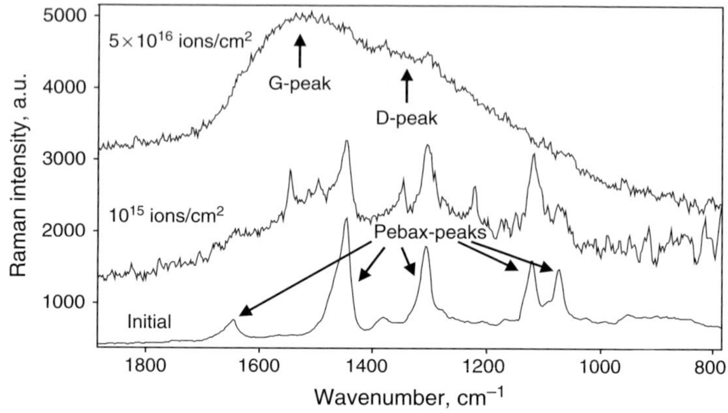

Fig. 7.1. Micro-Raman spectra of Pebax surface after PIII at 20 keV with different fluence.

fluence, the micro-Raman spectra of Pebax samples shows the same strong peak (Fig. 7.2). The decrease in intensity of the initial Pebax lines and the increase of the ordered graphite carbon peak are also observed. At high fluence treatment (from 10^{16} to 10^{17} ions/cm^2), the Raman spectra contain two sharp peaks at 1590 and 1360 cm^{-1}. At this treatment fluence, the spectral lines of unmodified Pebax

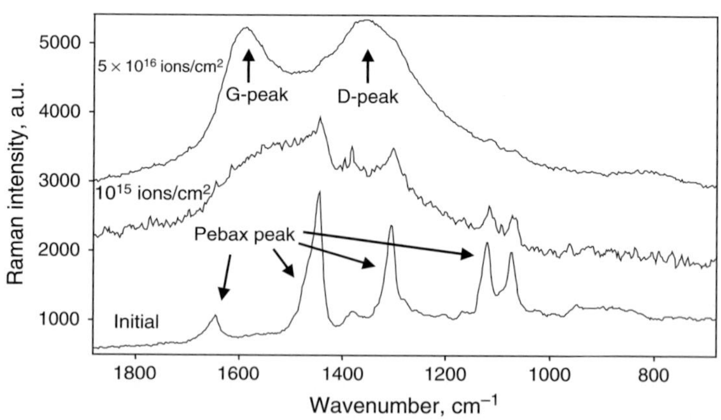

Fig. 7.2. Micro-Raman spectra of Pebax surface after PIII at 30 keV with different fluence.

macromelcules disappear completely. The fluence of 10^{16} ions/cm^2 corresponds to a stage of complete carbonization of the polymer surface layer under ion beam. Subsequent deposition occurs into the carbon layer, not into the polymer layer. At fluences higher then 10^{16} ions/cm^2, the carbon is created within the surface layer treated by nitrogen ions transformed into nano-crystalline graphitic structures containing high concentrations of crosslinking between graphite planes. Such structures have a high hardness.

The structure changes of the Pebax surface layer after ion beam implantation are observable by FTIR ATR spectroscopy. A new wide band appears in the 3600–3200 cm^{-1} wavenumber region in the spectra of treated Pebax (Fig. 7.3). This broad band contains the vibrational lines of hydroxyl and amine groups. In comparison to the narrow amide a line at 3306 cm^{-1} of the Pebax amide group, the new band is significantly wider and has some separated maxima. In the middle region of FTIR spectra the line of 1638 cm^{-1} and doublet of 1720/1737 cm^{-1} of Amide 1 are overlapped with a new broad band at 1650–1750 cm^{-1}. These new bands are interpreted as vibrations of carbon—carbon unsaturated groups and carbonyl groups. The new carbonyl,

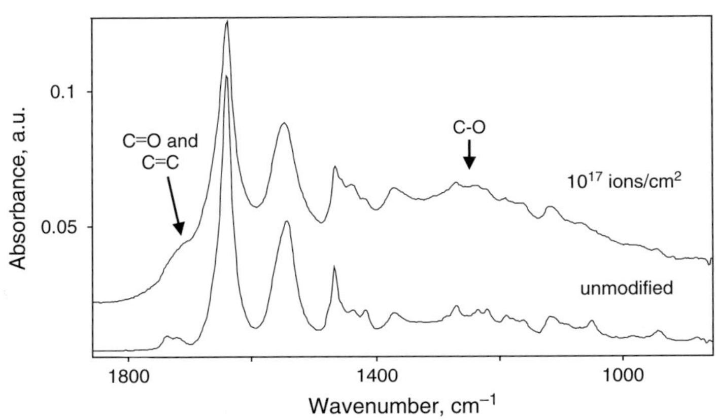

Fig. 7.3a. FTIR ATR spectra of Pebax after PIII of 30 keV energy nitrogen ions.

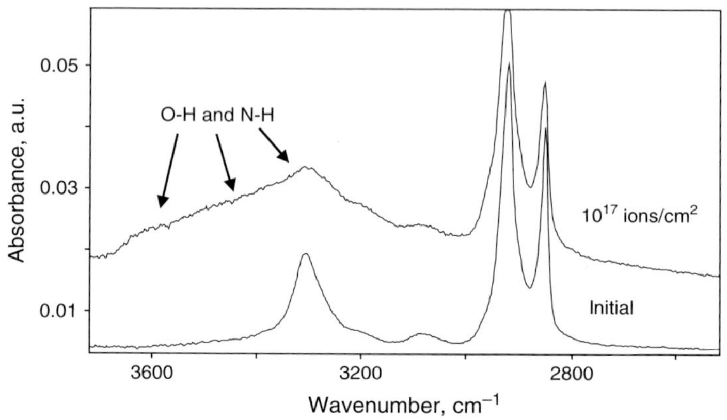

Fig. 7.3b. FTIR ATR spectra of Pebax after PIII of 30 keV energy nitrogen ions.

hydroxyl and amine groups appear as a result of depolymerization reactions in Pebax macromolecules generated by radicals produced under ion bombardment. In the process of depolymerization, the carbonyl, hydroxyl and amine groups are formed at the ends of broken macromolecules and a large number of groups with carbonyl, hydroxyl and amine fragments appear in treated Pebax. Such processes of depolymerization are also observed in polyethers and polyamides under high energy treatments like γ- and UV-light irradiations. In the case of ion beam implantation, the depolymerization process occurs at depths greater than the ion penetration depth, where free radicals from the highly carbonized layer propagate deep into the untreated region of the polymer and cause the reactions of depolymerization as it is observed under γ- and UV-light irradiations.

Based on the analysis of Pebax after ion beam implantation, the structure of modified polymer at high fluence of implantation could be presented as following Fig. 7.4. The top layer is highly carbonized. The structure of carbon depends on fluence and energy of ions. For high energy and high fluence, the graphite structures with high concentration of crosslinks between graphitic planes are formed. The thickness of this layer equals or is close to ion penetration

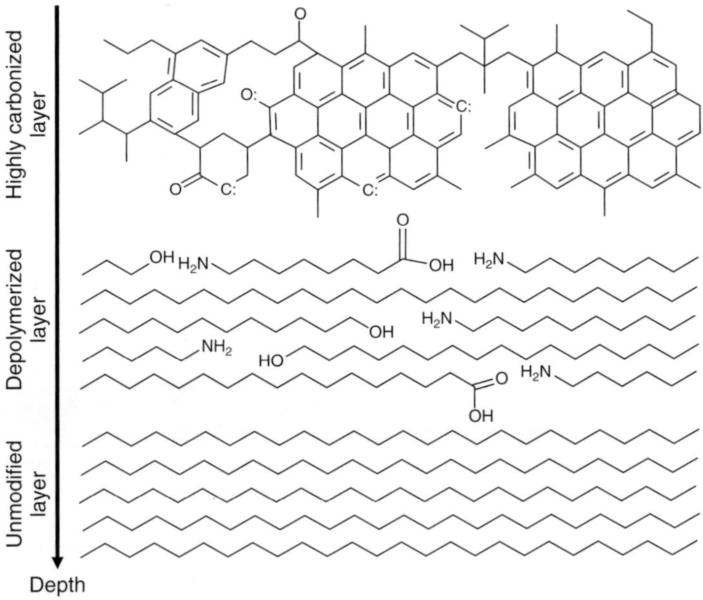

Fig. 7.4. Structure of Pebax surface layer after ion beam implantation.

depth. The next layer contains partly depolymerized and partly crosslinked polymer. Such a layer is thicker than the ion penetration depth. Its thickness depends on free radical propagation into bulk unmodified layers of polymer. We now consider the behavior of such a layered structure under indentation forces.

The hardness measurement of the thin surface layer was done by AFM method in contact mode. The curve for a silicone plate was used as a reference (Fig. 7.5). This curve shows a deformation of the cantilever in contact with a hard surface. The curve of the initial Pebax is less steep, corresponding to the deformation of the Pebax surface layer under the tip load. The penetration depth of the tip into the Pebax film equals the difference between the Silicon curve and the Pebax curve. At the maximum measurement point, the depth of penetration is about 50 nm. The asymptotic slope of the load curve is proportional to the modulus of elasticity of Pebax. The unloading curve shows

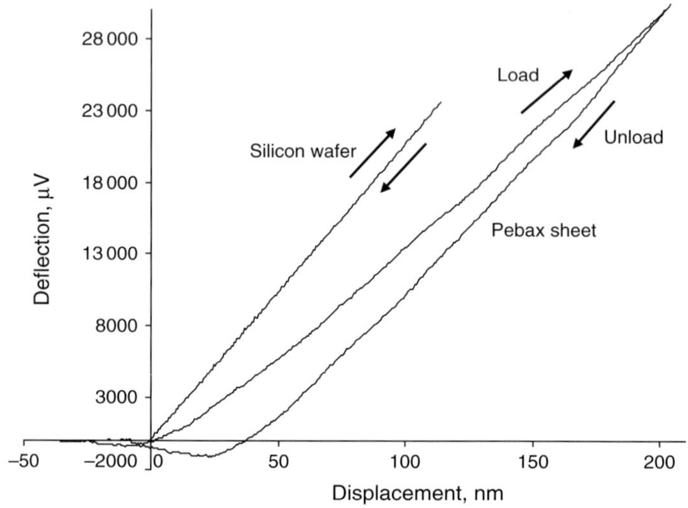

Fig. 7.5. AFM load curve for untreated Pebax sheet and silicon wafer (as reference). AFM measurements were done by P. Volodin.

that hysteresis occurs in the surface layer. This corresponds to mechanical reference data of Pebax at not destructive load [15]. The hysteresis is caused by mechanical energy loss due to movement and conformational transitions of polymer macromolecules under the load.

The curve of the ion beam modified Pebax has clearly a different form (Fig. 7.6). The asymptotic slope of the linear part of the curve is significantly steeper than for the initial Pebax plate. The unload curve is almost equal to the load curve. The hysteresis disappears. This behavior corresponds to a hard material with a low molecular mobility and a high modulus of elasticity. Taking into account the structure data at high treatment fluences, the surface layer of the treated Pebax film is a highly carbonized layer with a high modulus of elasticity.

At a mid range fluence of treatment, carbonization is not complete and the graphite structure is not hard in comparison with a high fluence. The depolymerization under the surface layer decreases its strength. Therefore, the polymer has the top surface layer with increased hardness and a

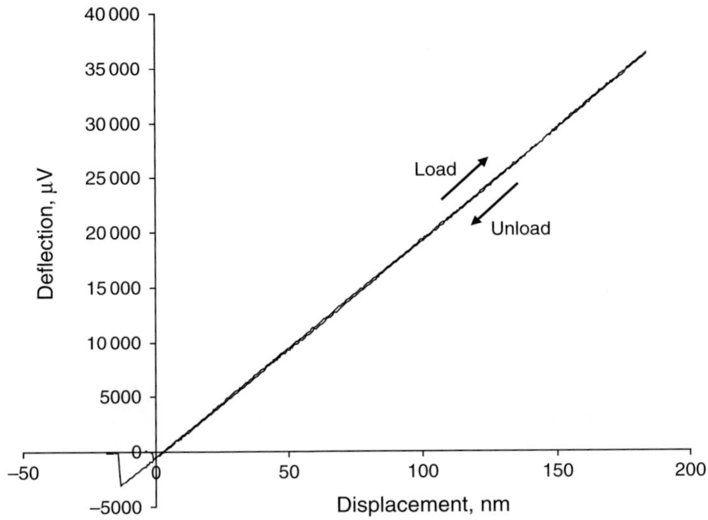

Fig. 7.6. AFM load and unload curve of Pebax plate after PIII of 30 keV energy and 10^{17} ions/cm^2 fluence. AFM measurements were done by P. Volodin.

second layer with decreased hardness relative to the initial polymer. The load curve shows two stages: at first, the tip penetrates into the hard layer, the hardness is quite close to that of a silicon plate; subsequently, the tip suddenly penetrates the soft layer, where stiffness is less than for the initial polymer (Fig. 7.7).

The elastic modulus of the polymer surface layer was calculated from AFM load curves [16]. The elastic module increases with ion beam implantation fluence and energy of ions (Fig. 7.8). The highest elastic modulus is observed for 30 keV energy of ions, at which a graphitic structure with a high concentration of crosslinks is observed. The elastic modulus of the surface layer increases relatively to ten times at ion beam treatment with a fluence of 10^{16} ions/cm^2 and energy of 30 keV in comparison with the untreated surface layer.

We observed the same curves for polyamide and different kinds of polyurethane sheets. Before modification, the load

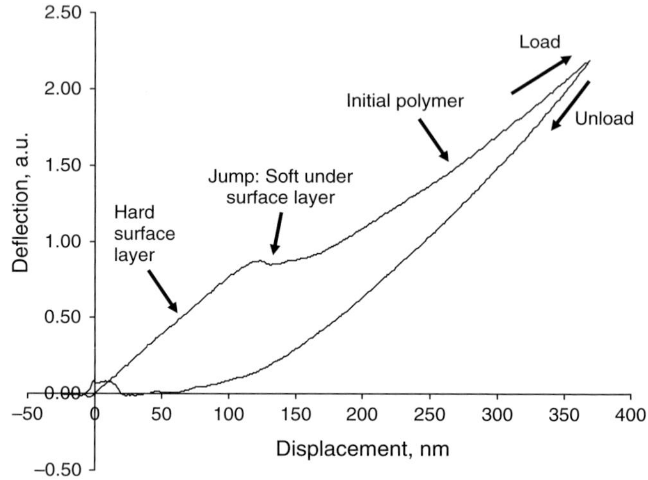

Fig. 7.7. AFM load and unload curve of Pebax plate after PIII of 30 keV energy with $5 \cdot 10^{15}$ ions/cm^2 fluence. AFM measurements were done by P. Volodin.

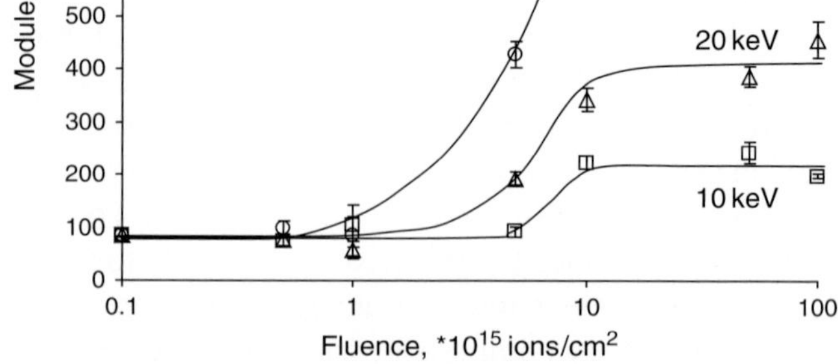

Fig. 7.8. Effective elastic modulus of Pebax sheet after PIII by AFM measurements. AFM measurements were done by P. Volodin.

curve is characteristic of soft material with large hysteresis (Fig. 7.9). After ion beam implantation at high fluence (up to 10^{16} ions/cm^2), the hysteresis disappears and the elastic modulus of the polymer surface layer increases (Fig. 7.10).

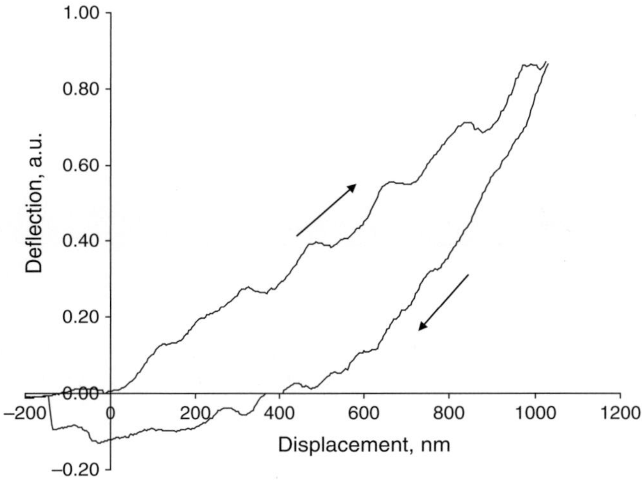

Fig. 7.9a. AFM load curve for untreated polyurethane sheet. AFM measurements were done by P. Volodin.

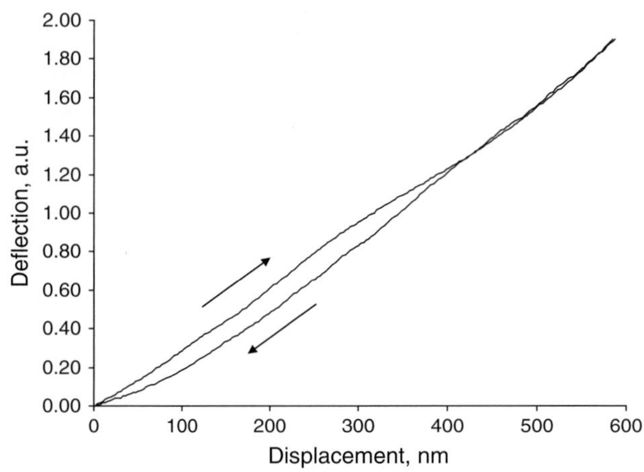

Fig. 7.9b. AFM load curve for PIII treated Polyurethane sheet. AFM measurements were done by P. Volodin.

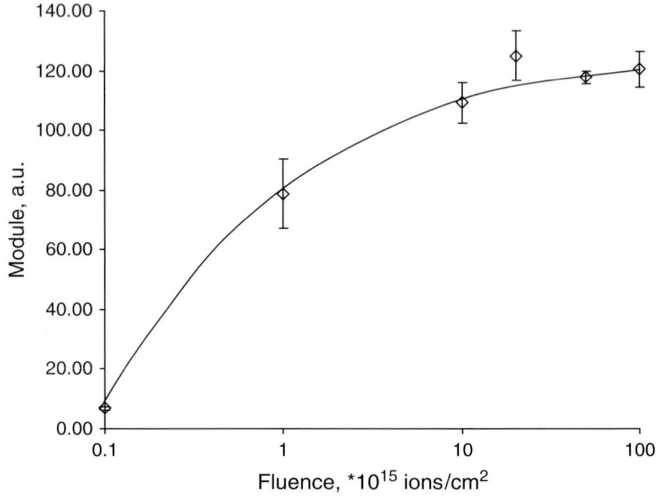

Fig. 7.10. Effective elastic modulus of polyurethane sheet after PIII by AFM measurements. AFM measurements were done by P. Volodin.

Some specific applications need a combination of specific mechanical properties with, for example, specific optical properties. We used ion beam implantation for improvement of polycarbonate and polyimide windows of aircraft which fly under conditions in which a high flux of dust particles occurs [17]. Such conditions exist in desert airports where frequent winds lift off dust clouds and aircraft take off with high speed in dense clouds of dust. The dust's abrasive action is weak but the ageing of the windows occurs significantly quicker due to the collection of scratches. This results in a decrease in optical transparency of the windows with time.

In our experiments, the plates of polycarbonate and polyimide of 2 mm thickness were treated with ion beam implanter with nitrogen ions of 40 keV energy. The ion beams in pulse and continuous regimes of implantation were applied. The current density did not exceed $20\,\mu A/cm^2$ in pulse and continuous regimes. A fluence of 10^{13}–10^{15} ions/cm^2 was applied.

The carbonization effect is observed by darkening of the polycarbonate and polyimide surface after ion beam implantation at high fluences (more than 10^{15} ions/cm^2). But at high fluence the transparency of the windows decreases due to a high degree of carbonization to levels not permitted for aircraft windows. The decrease of transparency starts from $5 \cdot 10^{14}$–10^{15} ions/cm^2, which is therefore an upper limit to the fluence that can be applied for window modification. In our case, the polycarbonate and polyimide windows were treated with a fluence of $2.5 \cdot 10^{14}$ ions/cm^2 which does not increase the light absorbance of windows in the visual region. Therefore, this fluence is suitable for window modification.

Scratch testing was carried out with a diamond indenter. The indenter was loaded with different weights and moved on a sample surface with constant speed. After scratching, the width and depth of the scratch were measured by optical microscopy. The experimental data are presented in Table 7.1. The indenter scratches the polymer with different depths corresponding to penetration of indenter in dependence on indenter load:

$$H_{scratch} = \frac{F}{d^2} \cdot k \tag{7.1}$$

Table 7.1. Depth of scratch and the scratch hardness of polymer surface after ion beam implantation: nitrogen ions of 40 keV energy with fluence of $2.5 \cdot 10^{14}$ ions/cm^2

Indenter load, g	Unmodified		Ion beam implanted	
	Depth of scratch, μm	$H_{scratch}$, kgf/mm^2	Depth of scratch, μm	$H_{scratch}$, kgf/mm^2
Polycarbonate				
100	42	57	48	43
25	21	57	21	57
5	8.5	78	2.3	945
Polyimide				
100	32	98	32	98
25	15	111	16	98
5	3.6	386	0.9	6173

where $H_{scratch}$ is the scratch hardness, F is the indenter load, d is depth of indenter penetration, and k is a coefficient of the indenter geometry.

At a high load of 100 g and 25 g the indenter penetrates through the carbonized layer. The depth of indenter penetration is 20–50 μm for polycarbonate and 15–30 μm for polyimide. It corresponds to a scratch hardness of about 60 kgf/mm^2 for polycarbonate and about 100 kgf/mm^2 for polyimide, which equals the scratch hardness of unmodified polymers. The indenter penetration depth is much higher than the thickness of the carbonized layer. The hardness improvement of surface layer cannot save the polymer from scratching at such high loads.

At a low load of 5 g the indenter penetrates less into the modified polymer than into the unmodified layer. But the depth of indenter penetration is still higher (200–900 nm) than the ion penetration depth at ion beam implantation (about 100 nm). The scratch hardness increases from 60 kgf/mm^2 to 945 kgf/mm^2 for polycarbonate and from 100 kgf/mm^2 to 125 000 kgf/mm^2 for polyimide.

Therefore, the ion beam implantation improves the scratch hardness of polycarbonate and polyimide surface layers. If exploitation occurs at low load, the carbonized surface layer could preserve the polymer against scratching. If exploitation occurs at high load, the carbonized layer will be damaged and the effect of ion beam implantation will not work. The ion beam modification is effective when the exploitation load is low enough. In such a case, the high hardness of the modified polymer corresponds to a highly carbonized structure, which is effective for long life polymer devices, for example for aircraft windows under dust atmosphere conditions.

Finally, we would like to note once more that the hardness improvement occurs in the thin surface layer of a polymer. If at mechanical load (scratching, wearing) the tip (or abrasive particles, and so on) penetrates through the modified layer, the effect of surface hardness does not work and the ion beam modification cannot improve hardness.

References

1. L. Calcagno, G. Compagnini, G. Foti, Structural modification of polymer films by ion irradiation, *Nucl. Instr. and Method.* B65 (1992) 413–422.
2. N. Gavrilov, D. Yakusheva, A. Kondyurin, Structure of polyethylene after pulse ion beam treatment, *J. Appl. Polym. Sci.* 69 (1998) 1071–1077.
3. M. Ueda, K.G. Kostov, A.F. Beloto, N.F. Leite, K.G. Grigorov, Surface modification of polyethylene terephthalate by plasma immersion ion implantation, *Surf. Coat. Technol.* 186 (2004) 295–298.
4. M. Guenther, G. Gerlach, G. Suchaneck, K. Sahre, K.-J. Eichhorn, B. Wolf, A. Deineka, L. Jastrabik, Ion-beam induced chemical and structural modification in polymers, *Surf. Coat. Technol.* 158–159 (2002) 108–113
5. E. Yap, D. G. McCulloch, D. R. McKenzie, M. V. Swain, L. S. Wielunski, R.A. Clissold, Modification of the mechanical and optical properties of a polycarbonate by 50 keV Ar and H ion implantation, *J. Appl. Phys.* 83(6) (1998).
6. R.S. Bhattacharya, Evaluation of high energy ion-implanted polycarbonate for eyewear applications, *Surf. Coat. Technol.* 103–104 (1998) 151–155.
7. F. Piazzaa, D. Grambole, D. Schneider, C. Casiraghi, A.C. Ferrari, J. Robertson, Protective diamond-like carbon coatings for future optical storage disks, *Diam. Rel. Mat.* 14 (2005) 994–999.
8. J.J. Rajesh, J. Bijwe, Investigations of scratch behaviour of various polyamides, *Wear* 259 (2005) 661–668.
9. C. Corbella, E. Pascual, M.A. Gomez, M.C. Polo, J. García-Cespedes, J.L. Andujar, E. Bertran, Characterization of diamond-like carbon thin films produced by pulsed-DC low pressure plasma monitored by a Langmuir probe in time-resolved mode, *Diam. Rel. Mat.* 14 (2005) 1062–1066.
10. R.C. Powles, D.R. McKenzie, N. Fujisawa, D.G. McCulloch, Production of amorphous carbon by plasma immersion ion implantation of polymers, *Diam. Rel. Mat.* 14 (2005) 1577–1582.
11. J.Y. Sze, B.K. Tay, Carbon ion implantation of ultra-high molecular weight polyethylene using filtered cathodic

vacuum arc with substrate pulse biasing, *Surf. Coat. Technol.* 200 (2006) 4104–4110.

12. A. Turos, J. Jagielski, A. Piatkowska, D. Bielinski, L. Slusarski, N.K. Madi, Ion beam modification of surface properties of polyethylene, *Vacuum* 70 (2003) 201–206.
13. W. Shi, X.Y. Li, H. Dong, Improved wear resistance of ultra-high molecular weight polyethylene by plasma immersion ion implantation, *Wear* 250 (2001) 544–552.
14. P.W. Serruys, M.J.B. Kutryk (Eds), *Handbook of Coronary Stents*, 2 edn, Martin Dunitz, London (1998).
15. A. Kondyurin, P. Volodin, J. Weber, Plasma immersion ion implantation of Pebax polymer, *Nucl. Instr. Meth. Phys. Res.* B 251 (2006) 407–412.
16. R. Prikryl, V. Cech, L. Zajickova, J. Vanek, S. Behzadi, F.R. Jones, Mechanical and optical properties of plasma-polymerized vinyltriethoxysilane, *Surf. Coat. Technol.* 200 (2005) 468–471.
17. N.V. Gavrilov, V.N. Mizgulin, R. Stinnett, A.V. Kondyurin, Modification of polymer films of PE, PTFE, PC, PI by pulse ion beams, *Khimicheskaya fizika i mesoskopiya* 1 (1999) 39–47.

8 Ion beam synthesis

Ion beam synthesis is a powerful method for the creation of new substances and specific layers in materials. Ion beam synthesis differs from ion beam modification due to its range of implanted fluences. The ion beam modification technique is used in the 10^{13}–10^{16} ions/cm^2 range of fluence. Such fluences are enough for the modification of initial materials; however, the concentration of implanted ions is not enough for the appearance of a new separate phase including implanted atoms. If fluence is higher than 10^{16} ions/cm^2 the separate phase with implanted ions (atoms, clusters) can be found in the target. Ion beam together with deposition processes are widely used for the synthesis of new structures in semiconductors and metals [1]. Ion beam synthesis in polymer materials could also be used for the creation of a new phase. In particular, it is useful for specific structures, which cannot be formed by other methods.

An example is the ion beam synthesis of metal nanoparticles in polymer surface layers [2]. The metal particles are created by implantation of metal ions into the polymer target. The energy used for implantation is in the range of 25–150 keV, which provides an ion penetration depth of 20–200 nm in the polymer target for heavy ions like Fe and Co. After ion penetration, the stopped metal atom is surrounded by polymer organic macromolecules. Due to high incompatibility with polymer macromolecules, the implanted metal atoms can diffuse and form metal clusters (particles). The process of particle formation could be divided at certain stages. First, the implanted atoms are collected to a critical portion, when the density of atoms in the polymer matrix becomes high enough for nucleation. At this stage, the nucleation of the implanted atoms gives small clusters (particles), which are joined to big particles.

Usually, the fluence of implantation used for metal particle synthesis is in the range of 10^{16}–5×10^{17} ions/cm^2. At fluences lower than 10^{16} ions/cm^2, the implanted ions are separated by a large amount of polymer and the diffusion of the atoms is not enough for nucleation. For example, at such fluences the iron atoms implanted into PMMA with an energy of 40 keV have about 5–10 monomer PMMA units between them. With a fluence increase up to 10^{16} ions/cm^2 the density of implanted ions is enough to form the metal clusters; however, the synthesized particles are rather small with a diameter of 1–2 nm. The distance between particles is too long. With fluence increase, the diameter of particles grows up to 20–30 nm and at fluences higher than 10^{17} ions/cm^2 the overlapping of metal clusters is observed. At fluences higher than 10^{17} ions/cm^2 the continuous metal film is formed on the polymer surface.

The process of particle formation depends on mobility of implanted ions in the polymer matrix. At high mobility the implanted atoms have the opportunity to diffuse a long distance from initial place after stopping. Therefore, the atoms can form clusters at lower concentrations in the polymer matrix if the diffusion of the atoms is higher.

The implanted ions cause structure transformation in the surface polymer layer at gaseous ion implantation. At high fluence of implantation the effect of carbonization is mostly observed in implanted polymers. Such carbon structures have low mobility and the diffusion of atoms in the implanted region of the polymer becomes harder with fluence of implantation. Additionally, active carbon atoms with dangled bonds form new bonds with implanting ions of metal. The formation of bonds between metal atoms and carbon of the surface layer prevents the diffusion of the metal atoms and the formation of metal clusters.

The mobility of the polymer matrix could be increased significantly if an implantation is performed into liquid polymer matrix with low viscosity [3–5]. Such implantation with high fluence was done for Fe, Co and Ag ions into uncured silicon rubber and epoxy resin matrixes which were in a viscous state during ion beam implantation.

For an implantation the liquid polymer matrix is placed in a vacuum chamber and implanted by energetic ions with high fluence. In the first case the polymer matrix remained liquid during whole process of implantation. In the second case, the matrix contained two active components, which reacted with the formation of solid polymer. The implantation starts when the matrix is in a viscous state. With time the matrix viscosity increases due to curing reaction up to complete solidification of the matrix. If the curing reaction is sufficiently slow, the ion beam implantation occurs in the polymer matrix at constant viscosity and the formation of the metal particles could be described in terms of hydrodynamic behavior of the metal clusters. After implantation and curing reaction the polymer matrix fixes the metal clusters and the implanted polymer target becomes stable.

The influence of viscosity on diffusion processes in the implanted matrix leads to differences in shape and size of metal particles implanted into the matrix in different states [4]. At the beginning, when the matrix has low viscosity, the implanting atoms have high mobility and the metal phase can easily move in the liquid matrix to form sufficiently large particles. In such a case, the shape of the particles can vary from needles to wormlike particles with long distances between particles. With viscosity increase, the distance between particles decreases, and the atoms do not have freedom to move long distances. However, the local mobility of implanting atoms gives the right shape of particles: spherical or cubic particles are formed depending on the conditions of implantation. In some cases, the formation of right form particles with highly ordered crystalline structure could be reached.

Synthesized metal particles could be used in different applications. For example, the magnetic materials with high magnetization could be created as basic material for storing magnetic devices. The optimal parameters of ion beam implantation depend on the viscous state of the polymer matrix. It gave a magnetization three times higher in films implanted in the liquid matrix than in films implanted in the solid matrix [5].

Another application is a non-linear optical device based on metal implanted polymers. Such a polymer device has gradient refractive index by thickness, which can be used in transformers of optical signals in integrated electro-optical elements. The combination of gradient refractive index with effect of surface Plasmon resonance on metal particles could be used in optical sensors integrated into non-linear optical elements.

However, the ion beam synthesis of metal inclusions in the liquid polymer matrix needs follow-up investigation and development which can bring new physical effects and applications.

The ion beam implantation by gas ions with high fluence of implantation could be used for the generation of a new phase in the implanted surface layer of polymers. The final structure of the implanted target depends on initial atomic contamination of the polymer target and on implanting ions. For example, the ion beam implantation of high energy ions into silicone rubber [6, 7] generates a mixture of silicone oxide and silicon carbide layer on the top of the rubber. We have observed the appearance of such a silicon carbide phase by Si-C band [8] in spectra of silicone rubber after ion beam implantation with argon ions (Fig. 8.1). In such case, the ion beam implantation could be used as a method of ion beam synthesis for inorganic layers on the top of the polymer surface.

The simplest case is ion beam implantation into hydrocarbon polymers, which generates a completely carbonized layer on the top of the polymer or any other substrate. If the thickness of the polymer is low enough and the penetration depth of implanting ions is high enough, a complete carbonized film from organic polymer can be created.

For example, a polystyrene thin layer was used as initial polymer target on the surface of silicon wafer [9, 10]. The polystyrene film with a thickness of 100 nm was spincoated

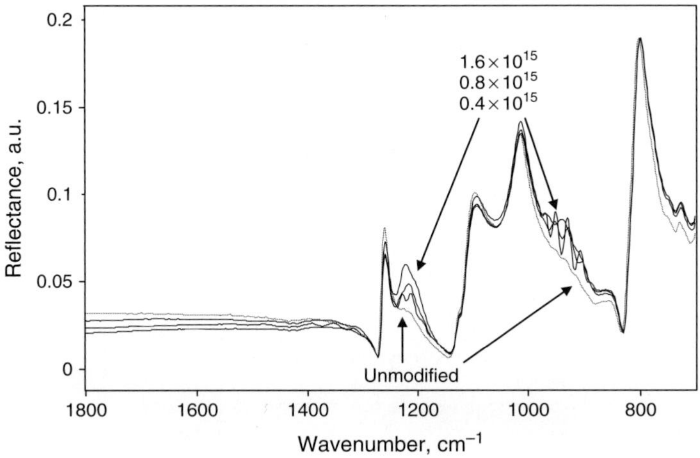

Fig. 8.1. FTIR reflection spectra of silicone rubber after ion beam implantation. Fluence is noted in ions/cm^2.

on the silicon wafer. The topology of the polystyrene film repeats completely the topology of the silicon wafer (Fig. 8.2). The ion beam implantation was done with nitrogen and argon ions with high energy of 20 keV and high fluence of 10^{16} ions/cm^2. After ion beam implantation the topology of the surface remains, and the roughness parameter RMS does not change much (from 0.246 for spincoated polystyrene to 0.247 nm for modified polystyrene). The FTIR spectrum of polystyrene film shows the decrease of polystyrene line intensity with fluence of ion beam implantation (Fig. 8.3). The decrease occurs due to etching of polystyrene film under ion beam and due to structure transformation of polystyrene into carbon film. Etching decreases the thickness of polystyrene film. Etching at low energy of ions (plasma discharge) is used in the electronic chip industry for the removal of photoresists. However, in our case, the etching process occurs with high rate only at low fluence of implantation, when the film has polymer structure and the penetrating ions cause dehydration and spattering in the polymer film. At high fluence of implantation, the etching process becomes slow due to a high degree of

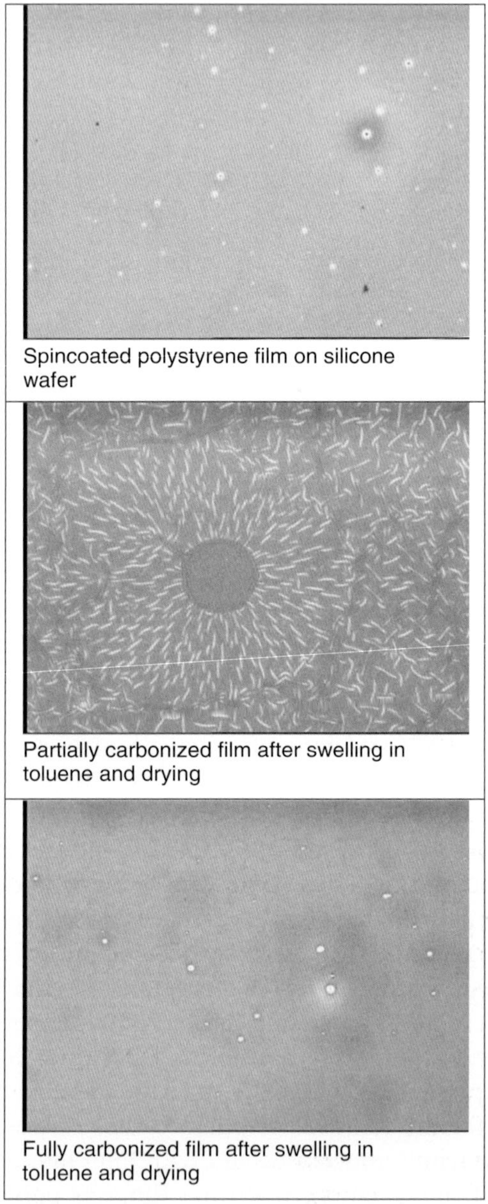

Fig. 8.2. Optical microscope image of polystyrene film on silicon wafer before and after ion beam implantation. Image size is 1 × 1.2 mm for 1st and 3rd pictures and 20 × 24 μm for 2nd picture.

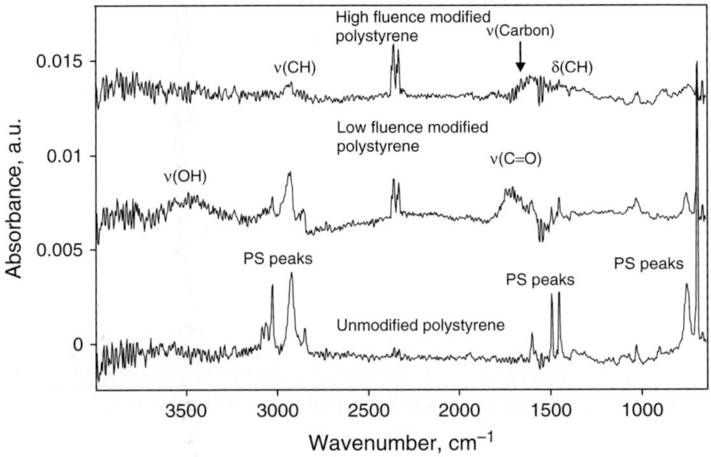

Fig. 8.3. FTIR transmission spectra of polystyrene film on silicon wafer (spectra of silicon is subtracted) before and after ion beam implantation.

carbonization of the film. The strong carbonization effect of polystyrene under ion beam does not permit a complete etching of the film. For example, 100 nm polystyrene film is etched up to 40–60 nm completely carbonized film on silicon and gold substrates at high fluence of implantation (10^{16}–5×10^{16} ions/cm^2).

The structure transformations in polystyrene film are observed by FTIR transmission spectra. After ion beam implantation the lines attributed to vibration of aromatic groups in polystyrene macromolecules disappear completely. The calculated residual hydrocarbons in the form of $=$CH$-$ and $-CH_2-$ groups remain about 10–15 % in highly implanted film. The FTIR spectrum of high fluence modified film also shows the presence of carbon structure by line in the 1600 cm^{-1} region of spectra. By ellipsometrical measurement data, the refractive index of ion beam implanted polystyrene film in the 500 nm wavelength region of light increases from 1.6 to 2.1–2.3 values, which corresponds to carbon materials like graphite, diamond or its mixture.

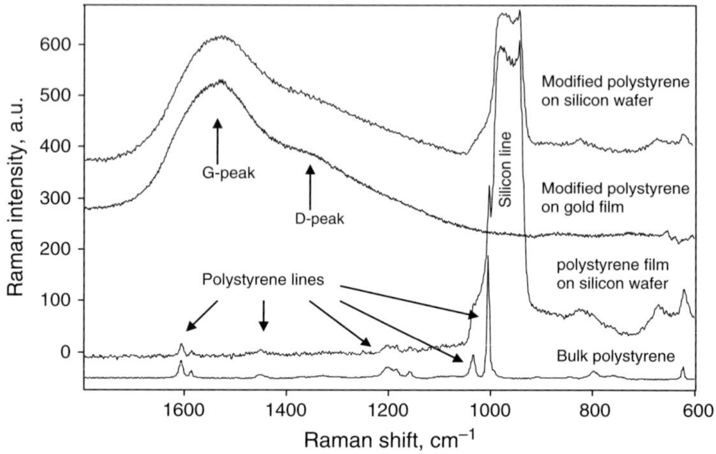

Fig. 8.4. Micro-Raman spectra of polystyrene nano-film on silicon and gold substrates before and after ion beam implantation with high fluence.

The completely carbonized layer on the top of silicon wafer after ion beam implantation into polystyrene film is observed by micro-Raman spectra (Fig. 8.4). The spectra of unmodified polystyrene (bulk and on silicon wafer) show narrow lines attributed to vibrational modes of polystyrene macromolecules. After ion beam implantation into polystyrene, the Raman spectra show lines of carbon structures: G- and D-peaks. Detailed consideration of such peaks is presented in Chapter 4. The presence of these peaks corresponds to amorphous carbon phase with sp^2 and sp^3 hybridized carbon atoms. The film becomes completely insoluble in any organic solvent and stable at high temperatures (Fig. 8.2).

Therefore, after PIII treatment the polystyrene film was completely transformed into carbon coating on silicon wafer. After carbonization the film could be removed from the silicon wafer by floating in base or acid media to dissolve silicon oxide under the carbon layer. The film is sufficiently durable in the free state and can be placed on a TEM grid or on a wide pore membrane. For example, the carbon films of some mm^2 area and 10–30 nm thickness or thicker could be synthesized this way.

Such synthesis for inorganic coatings and membranes using ion beam implantation is suitable when the polymer film could not be annealed up to the formation of pyrolytic carbon. Also the ion beam synthesis could be used for polymers, which are decomposed under high temperature (for example, PMMA and PLA) and the pyrolytic method could not be applied. The ion beam implantation transforms such kinds of polymers into pure carbon structure.

The ion beam synthesis from polymer structures can play a significant role in synthesis of complex shape inorganic nano-structures. For example, highly carbonized nano-membranes based on block-co-polymer self-ordered film with pores of 10 nm diameter are synthesized using ion beam. In spite of this, different applications can be found for ion beam synthesis of dewetted polymer and also organic structures. This field of ion beam application for polymers has not yet been well investigated and developed despite a wide prospective.

References

1. H. Weishart, V. Heera, F. Eichhorn, B. Pecz, A. Barna, W. Skorupa, Ion beam synthesis of diamond-SiC-heterostructures, *Diam. Rel. Mat.* 12 (2003) 1241–1245.
2. A.L. Stepanov, Optical extinction of metal nanoparticles synthesized in polymer by ion implantation, in *Metal-Polymer Nanocomposites*, edited by Luigi Nicolais and Gianfranco Carotenuto, John Wiley and Sons, 2005.
3. S.N. Abdullin, A.L. Stepanov, Yu.N. Osin, R.I. Khaibullin, I.B. Khaibullin, Synthesis of metallic dispersion and continuous films in the viscous polymer by implantation of cobalt ions, *Surf. Coat. Technol.* 106 (1998) 214–219.
4. I.B. Khaibullin, R.I. Khaibullin, S.N. Abdullin, A.L. Stepanov, Yu.N. Osin, V.V. Bazarov, S.P. Kuzin, Ion metal synthesis in viscous organic matter, *Nucl. Instr. Meth. Phys. Res.* B 127/128 (1997) 685–690.

5. R.I. Khaibullin, B.Z. Rameev, V.N. Popok, E.P. Zheglov, A.V. Kondyurin, V.A. Zhikharev, B. Aktas, An influence of the viscosity of polymer substrate on ion beam synthesis of iron granular films, *Nucl. Instr. Meth. Phys. Res.* B 206 (2003) 1115–1119.
6. T.G. Vladkova, I.L. Keranov, P.D. Dineff, S.Y. Youroukov, I.A. Avramova, N. Krasteva, G.P. Altankov, Plasma based Ar^+ beam assisted poly(dimethylsiloxane) surface modification, *Nucl. Instr. Meth. Phys. Res.* B 236 (2005) 552–562.
7. G. Marletta, A. Toth, I. Bertoti, T.M. Duc, F. Sommer, K. Ferencz, Optical properties of ceramic-like layers obtained by low energy ion beam irradiation of polysiloxane films, *Nucl. Instr. Meth. Phys. Res.* B 141 (1998) 684–692.
8. E. Elmer, W.S. Ptak, FTIR studies of structural effects due to boron addition in sintered silicon carbide, *Vib. Spectros.* 29 (2002) 211–215.
9. A. Kondyurin, B.K. Gan, M.M.M. Bilek, K. Mizuno, D.R. McKenzie, Etching and structural changes of polystyrene films during plasma immersion ion implantation from argon plasma, *Nucl. Instr. Meth. Phys. Res.* B 251 (2006) 413–418.
10. B.K. Gan, M.M.M. Bilek, A. Kondyurin, K. Mizuno, D.R. McKenzie, Etching and structural changes in nitrogen plasma immersion ion implanted polystyrene films, *Nucl. Instr. Meth. Phys. Res.* B 247 (2006) 254–260.

9 Biological and medical applications

Modern polymers have a vast application in biology and medicine. Most medical devices are made from polymer materials, for example supporting devices for patients, storing and treatment devices for medications, physiological substances and organs, and prosthetics of different applications including artificial organs, skin and extremities. Replacements for human and animal body parts made from polymer materials must have high biocompatibility. The polymer must induce a low inflammatory reaction from the organism and have: low toxicity, similar elastic properties to biological tissue; and a long lifetime in organism media with satisfactory mechanical strength. There has been rapid growth in the use of polymer devices in medical applications over the last few years. However, despite this success there are still several major problems that need to be solved before the application is perfect for use in the human body. The modification of a polymer surface using ion beam implantation can be advantageous in biological and medical applications. Here we consider a few applications of ion beam implantation for medical polymers, which were found during our scientific projects and contracts with medical companies.

The attachment of proteins to polymer surfaces occurs immediately after inserting a polymer prosthetic into a living organism. This stimulates a secondary response by the organism's immune system. There are also applications of surface attached enzymes in the food, chemical, biotechnology and medical diagnostic industries. An example is a biosensor (biochip) based on enzyme activity for detection of antibodies in media for testing blood serum [1]. A biochip has to contain attached protein dots distributed on the polymer substrate. Proteins have to be active on the polymer, to be fixed on their position and to be read after soaking in analytical solution. Therefore, the polymer surface must be attractive for proteins and strong enough for proteins to adhere to it and save their biological properties (conformation, activity) as initial.

We used the enzyme horst radish peroxidaze (HRP) and bovine liver catalase for attachment to polyethylene, polystyrene and polytetrafluorethylene [2–4].

HRP and catalase are proteins with highly efficient enzyme properties. The molecular weights of catalase and HRP are 57 kDa and 44 kDa, respectively. Catalase and HRP protect cells from the toxic effects of hydrogen peroxide and catalyse the hydrolysis of hydrogen peroxide into oxygen and water according to the reaction

$$2H_2O_2 \leftrightarrow O_2 + 2H_2O \qquad (9.1)$$

Catalitic activity is provided by the presence of a porphyrin ring in the protein molecule. The catalytic activity depends strongly on conformation of the protein molecule. The activity can be lost at temperature, pH or chemical agent disturbance. The polymer surface can also influence protein conformation due to intermolecular interactions between protein molecule and polymer surface.

A protein molecule has hydrophobic and hydrophilic tails and can therefore attach to untreated hydrophobic polymer surfaces. The protein molecule is enclosed in a water sheath with hydrophilic tails turned to the outside of the protein space constituting to the hydrophilic properties of the protein. The virgin polymer macromolecules and protein hydrophobic groups interact via Van-der-Waals forces. On contact with a hydrophilic polymer surface, the conformation of the protein (which typically resides in a state suitable for a water environment) changes and the hydrophobic tails interact with the polymer macromolecules. In this case, protein is unfolding and conformational changes are expected.

We put the HRP molecule on a smooth polystyrene surface spincoated on silicon wafer. The dimensions of the HRP molecule are $3.5 \times 6.0 \times 7.5\,\text{nm}$ [5, 6]. Figure 9.1 shows the AFM images for an untreated PS surface with attached HRP layer. One can see hills distributed on the polystyrene surface. The surface morphology was analyzed based on quantitative analysis of AFM experimental data. The roughness of the surface is characterized by root mean square (RMS) as a parameter of deviation of surface pixels from the medium surface level:

$$R = \sqrt{\frac{1}{N^2} \cdot \sum_{i=1}^{N} \sum_{j=1}^{M} (H(i,j) - \overline{H})^2} \tag{9.1}$$

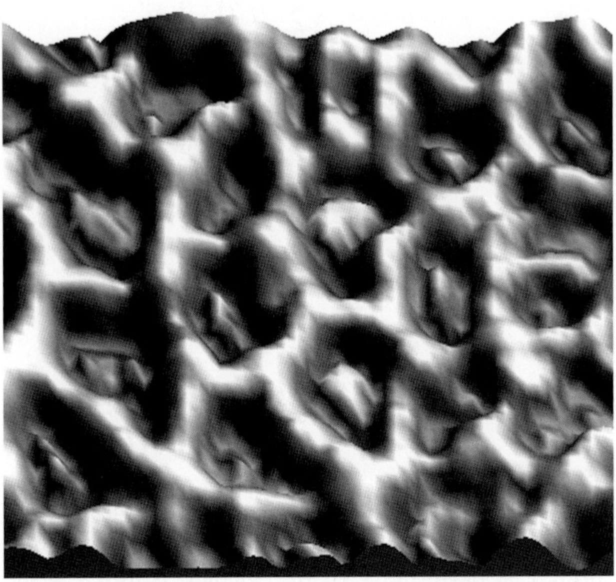

Fig. 9.1. AFM image of HRP coating on untreated polystyrene surface. Size of image is $400 \times 400\,\text{nm}^2$. Height of hills is 5 nm.

where H is the surface pixel level, I and j are pixel locations, and $\overline{H}$ is the medium surface level.

The RMS roughness parameter for the untreated spin-coated PS film is 0.78 nm, which corresponds to the RMS parameter of the initial silicon wafer. The RMS roughness of the surface after HRP attachment is 4.3 nm. The RMS parameter characterizes average deviation of the surface pixels without detailed analysis of lateral spacing of various features. The detailed analysis of the histogram of roughness events for HRP attached polystyrene surface gives an 8 nm hill height. Together with the AFM image it can be interpreted that the height of the protein hills is approximately 8–10 nm, similar to the height of a single HRP molecule (7.5 nm) [5].

Another useful roughness parameter is the power spectral density (PSD) which calculates roughness amplitude squared as a function of spatial wavelength of the features:

$$PSD(k) = \frac{1}{L^4} \sum_{n=0}^{N} \left| H(k_x, k_y) \right|^2 \tag{9.2}$$

where L is the size of the AFM image and $H(k_x, k_y)$ is the result of the fast Hartley transformation:

$$\begin{aligned} H(k_x, k_y) = {} & \left(\frac{L}{N}\right)^2 \sum_{x=0}^{N-1} \sum_{y=0}^{N-1} z(x, y) \\ & \cdot \left\{ \cos\left[2\pi\left(\frac{k_x \cdot x}{N} + \frac{k_y \cdot y}{N}\right)\right] \right. \\ & \left. + \sin\left[2\pi\left(\frac{k_x \cdot x}{N} + \frac{k_y \cdot y}{N}\right)\right] \right\} \end{aligned} \tag{9.3}$$

In these terms PSD is a Fourier decomposition of the image into special frequencies, and features of the PSD curve are interpreted as periodicity of the surface features with definite dimensions. The PSD parameter for lateral analysis of HRP islands on untreated polystyrene gives a peak at 92 nm, which characterizes the average distance between islands. Assuming that protein islands have a height less than 10 nm we also estimated the percentage coverage of hills over a height of 10 nm. The percentage coverage of hills is approximately 27% of the whole surface. Using these contours, the first neighbor distance (FND) as average minimal distance between hills was calculated. For HRP hills of 10 nm we obtained a value of 64 nm for FND. A combination of AFM, ellipsometry and FTIR results indicates that the HRP protein molecules form islands with heights of around 8 nm (corresponding to the longest dimension of the protein molecule) and lateral sizes of around 30 nm on the untreated polymer surface (Fig. 9.2) with a distance between islands of 64 nm.

In contrast with the untreated surface the PIII modified surface of polystyrene is completely covered with HRP

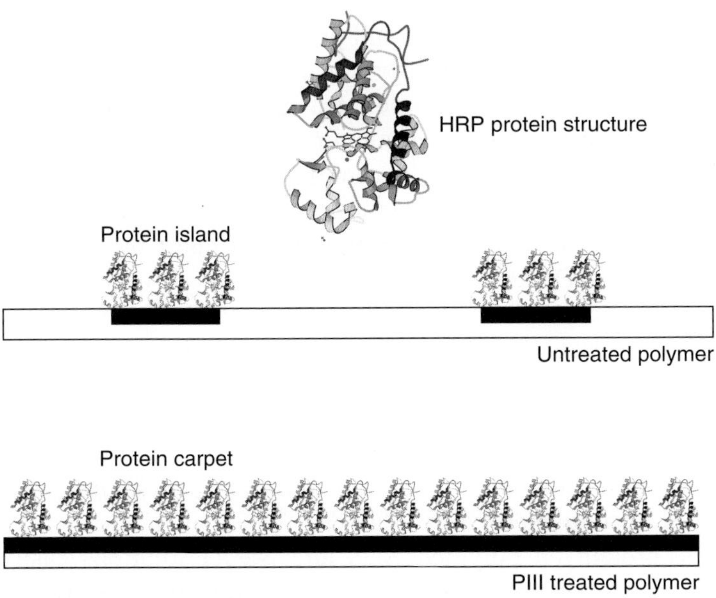

Fig. 9.2. HRP protein molecules distributed on untreated and PIII treated polymer surface. HRP protein molecules form islands on untreated polymer and continuous carpet on modified polymer.

molecules, like a carpet. The AFM, FTIR and ellipsometrical analyses show that 100% of the surface is covered with a uniform monolayer of HRP. The thickness of the covering equals the longest dimension of the protein molecule. Such vertical positioning of the HRP molecule reduces the contact area with polystyrene substrate, saves maximal contact with water media and covers maximally the polystyrene surface with protein molecules.

On a modified polymer surface the protein molecule has a number of intermolecular interaction possibilities. For example, aromatic condensed structures as acceptors of protons attract polar groups of molecules with flexible protons such as carboxyl groups, hydroxyl groups, and amine groups. The HRP molecule contains such groups sufficient for donor/acceptor interactions. There are a wide number of active groups: carboxyl, aldehyde, hydroxyl, peroxide as well as free radical groups, which can react with active groups of protein to form covalent bonds between polymer surface molecular groups and protein molecules. Hydrogen acceptor and donor groups are presented in the modified polymer surface to be included also in hydrogen bonds between surface molecules and protein molecules.

To determine the mechanism that makes the protein stick to the modified polymer surface we tried to remove the attached HRP and catalase protein molecules by chemical detergent SDS and NaOH solutions. Under active detergents protein molecules lose biologically active conformations and become simple inactive polyamide macromolecules. Such macromolecules cannot be saved on inactive substrate without chemical bonds after washing in proper solvents.

In reality, proteins were completely removed from the untreated polymer surface. If the surface of the polymer is untreated, only Van-der-Waals interactions hold the protein on polymer. SDS and NaOH solutions easily removed the protein layer from polyethylene, polystyrene and polytetrafluorethylene surfaces as viewed, for example, by FTIR ATR spectra of the polyethylene surface in Fig. 9.3. The presence of protein is observed by specific

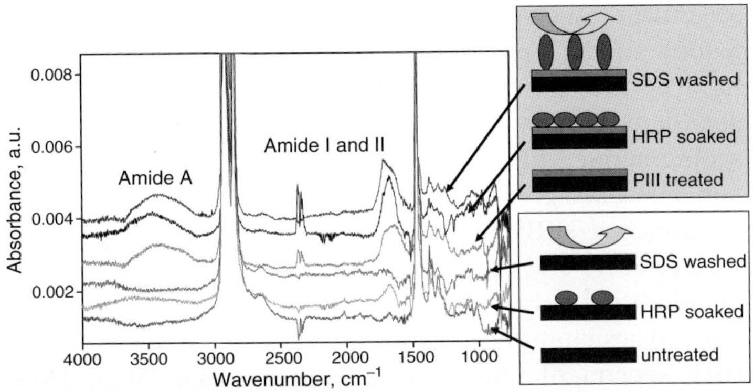

Fig. 9.3. FTIR ATR spectra of ultra-high molecular mass polyethylene (UHMWPE) after PIII, soaking in HRP protein solution and washing in SDS detergent: untreated UHMWPE, after HRP, after HRP and SDS, after PIII, after PIII and HRP, after PIII, HRP and SDS. Experiments were done in cooperation with N.J. Nosworthy.

spectral lines of protein: Amide A, Amide I and Amide II lines. If the surface is untreated, the protein sticks to the surface less than on the treated surface. Also FTIR spectra do not contain any lines of protein after washing the untreated polyethylene surface with attached protein in SDS or NaOH solution. If the surface is modified by ion beam, the protein cannot be removed by SDS solution even at high temperature (70°C). Protein molecules are jointed covalently with the polyethylene surface and only destruction of the polyethylene surface layer can remove protein molecules from the substrate.

Conformation of protein can be analyzed by the Amide I line of the FTIR spectra. We analyzed the conformation of HRP molecules attached to the polyethylene surface. The Amide I line is formed by separated individual lines related to vibrations of the amide group in the backbone of the protein molecule. Each line corresponds to combinations of C=O stretch vibration (main contribution, about 40–60% of energy), N—H deformation vibration (20–30% of energy) and some complex vibrations of bonds and angles in the amide group [7]. Such group vibrations are sensitive to the formation of hydrogen bonds and conformation of the protein molecule. Usually, the position of lines is different for vibrations in α-helixes, β-sheets and disordered chains of the protein molecule. The fitting procedure applied to the Amide I line based on literature and our own data gives the presence of all such structures in the attached HRP molecule on the untreated polyethylene surface (Fig. 9.4). The conformation of HRP molecule on the modified polyethylene surface is changed: lines of disordered structures become more intense in comparison with HRP spectra of untreated polyethylene (Fig. 9.5). However, the same region of spectra is related to vibrations of carbonyl groups in the surface layer of polyethylene. These carbonyl groups appear as a result of polyethylene surface layer oxidation and complete subtraction of the same oxidized layer spectra cannot be achieved. The spectra of the SDS treated surface show a sharp increase in the intensity of the disordering structure lines and a decrease in the intensity of α-helixes' and β-sheets' lines

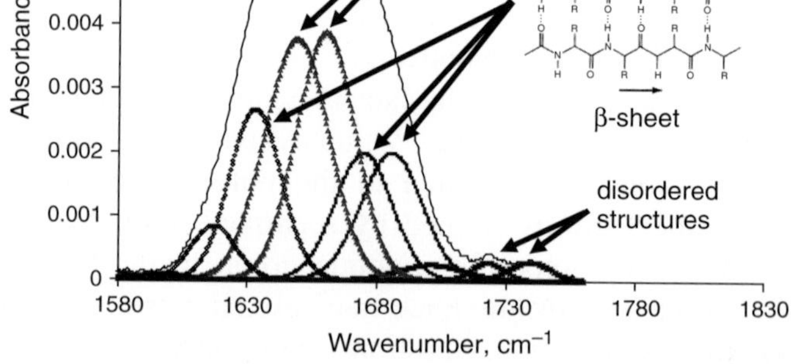

Fig. 9.4. Amide I line (top) and its components (bottom) in differential FTIR ATR spectra of untreated polyethylene with attached HRP protein molecules.

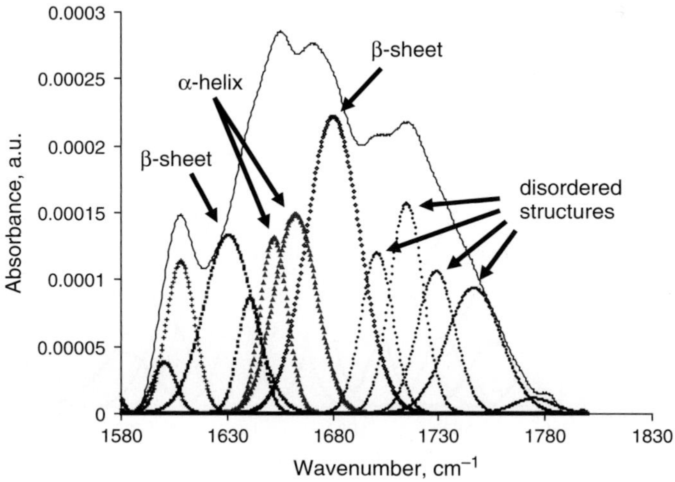

Fig. 9.5. Amide I line (top) and its components (bottom) in differential FTIR ATR spectra of PIII treated polyethylene with attached HRP protein molecules.

of HRP molecules (Fig. 9.6). The strong increase in the intensity of disordered structure lines for soaked HRP molecules supports the fact that protein molecules are unfolded and lose virgin conformation. Despite this loss, the protein molecules are still jointed to the polyethylene surface. Disordered structures are observed in attached HRP molecules on the modified polyethylene surface before using detergents, indicating that some molecules lose their virgin conformation on attachment to the modified polymer surface.

If a protein molecule is bonded covalently, there is a danger that the protein molecule is less active, because the conformation of the protein molecule has changed. The test for protein activity was done based on the well-known hydrogen peroxide reaction fixed by color TMB test. Hydrogen peroxide of 75 μl volume and 6 mM concentration was added to the polymer surface and incubated for 6 minutes by shaking. After incubation, 3 μl was removed and the remaining peroxide was assayed by a modified method of

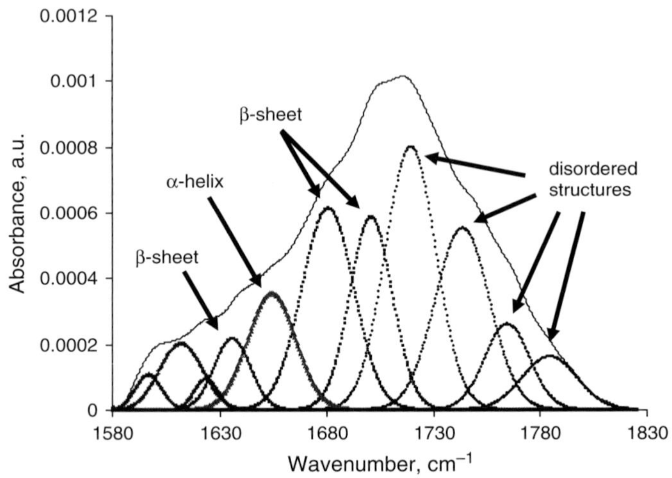

Fig. 9.6. Amide I line (top) and its components (bottom) in differential FTIR ATR spectra of PIII treated polyethylene with attached HRP protein molecules after washing by SDS and NaOH solutions. Experiments were done in cooperation with N.J. Nosworthy.

Cohen [8]. The hydrogen peroxide was added to 0.25 ml of a solution consisting of a mixture of 0.6 N H_2SO_4 and 10 mM $FeSO_4$, and 20 μl of 2.5 M KSCN was added to develop the color. Absorbance was measured at 475 nm using a DU 530 Beckman spectrophotometer. The absorbance shows an amount of active HRP molecules presented in the solution. Because before test the polymer was washed six times in buffer without protein, the activity can be attributed only to protein molecules sticking to the polymer surface. The activity of protein increases after ion beam implantation from 0.2 to 0.7 of TMB test absorbance for the polyethylene surface and from 0.07 to 0.56 for the polytetrafluorethylene surface. The amount of active stuck protein does not depend on fluence of treatment (Fig. 9.7). The same result is observed by intensity of protein lines in the FTIR ATR spectra of polymers. For example, the dependence of Amide II and Amide A line intensity shows a step increase after ion beam implantation of thr polytetrafluorethylene surface and fluence increase does not affect the amount of

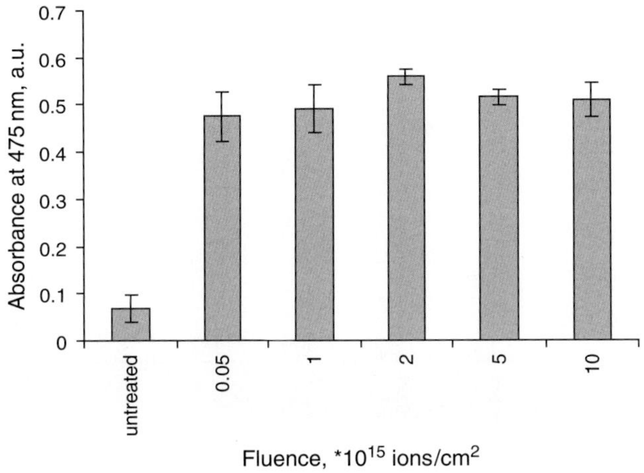

Fig. 9.7. Result of TMB array test for active HRP molecules attached to polytetrafluorethylene surface after PIII.

stuck protein (Fig. 9.8). Interestingly, the activity of the polymer surface to accept protein remains for months after ion beam implantation, if the modified polymer is stored under normal laboratory conditions. Thus, the ion beam modified polymer surfaces can easily attach enzymes and the enzymes remain functional.

The universal character of protein attachment after ion beam modification is supported by experiments with HRP, soyabean peroxidaze, catalase and elastan proteins. If the ion beam is somehow focused or the ion beam implantation is done with a mask, the protein may be attached in a pattern. For example, a dot pattern used in biochip technology can be generated: protein molecules can be easily removed from untreated fields using a detergent and the remaining molecules are attached to the implanted areas. Therefore, a polymer substrate with dots of stuck proteins can be prepared for biochip technology.

However, the ability of the modified polymer surface to attach proteins can be used for improvement of cell attachment. Cell attachment occurs via integrin, collagen, fibronectin, laminin, and transmembrane receptors

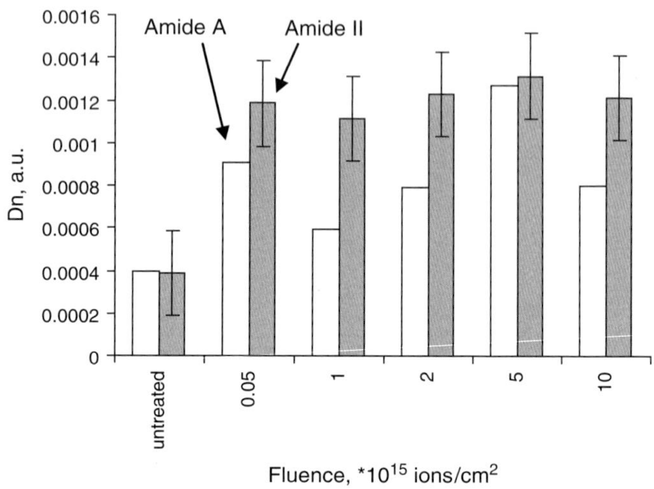

Fig. 9.8. Normalized absorbance of attached HRP lines in FTIR ATR spectra of polytetrafluorethylene after PIII: gray bars – 1540 cm^{-1} (Amide II) line, white bars – 3315 cm^{-1} (Amide A) line. Experiments were done in cooperation with N.J. Nosworthy.

at proteins of the extracellular matrix (ECM) and their basal membrane. Therefore, if the polymer surface becomes active for protein attachment, the surface could be active for cell attachment.

The adhesion of cells to artificial polymer substrates is of great significance for: growing cell cultures in artificial media; cell manipulation in bionic devices; growing artificial tissues; and for human and animal implants. The attached cell must be living, have good adhesion, and produce new cells (mitosis).

The human endothelial cell line HMEC-1 was attached to polyethylene and polytetrafluorethylene after ion beam implantation. The cells were seeded on the samples with a concentration of 5×10^4 cells per cm^2. The samples with cells were stored in an incubator at 36.7°C for 2 days. After that, an Alamar Blue™ test was performed for 4, 6 and 8 hours. In another variant, the seeding cells of the samples were fixed in 0.2% paraformaldehyde in PBS and stained with phalloidin-TRITC and DAPI for polymerized actin of the cytoskeleton (red) and cell nuclei (blue). Images were taken by fluorescent microscopy and cell distribution was calculated.

The images of cells on the polytetrafluorethylene surface after 2 days of growing are presented on Fig. 9.9. On the untreated surface, the cells form islands with the high density of cells. Such an image is quite close to the initial distribution of cells at seeding. On an ion beam modified surface the cells are uniformly distributed over the whole area of the sample. The number of cells on the untreated surface does not change after growing for 2 days. The number of cells is three times greater on the modified surface than on the untreated surface (Fig. 9.10). This means that the cells grow on the ion beam modified surface.

The quantitative analysis of cell distribution can be done by density of cell distribution calculated as the n_{ij} number

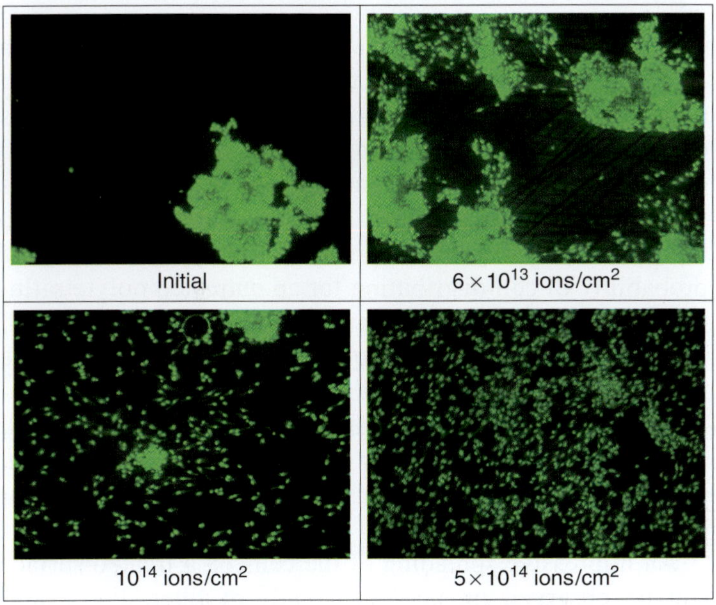

Fig. 9.9. Microphotos of cells on surface of PTFE film treated with ion beam with different fluences. Experiments were done in cooperation with M. Maitz and I. Kondyurina.

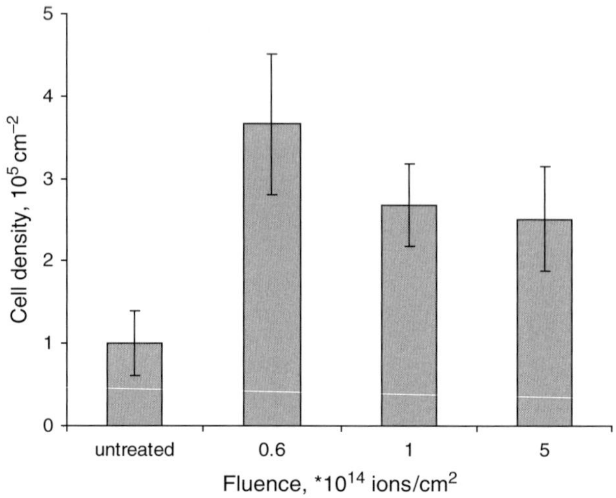

Fig. 9.10. Cell density on polytetrafluorethylene surface modified by ion beam after 2 days of cell growth. The experiment was done in cooperation with I. Kondyurina and M. Maitz.

of cells seeding in the s_{ij} unit area. A number of areas with n_{ij} cells can be interpreted as the probability of finding such a density of cell distribution in a random area. The results of such analysis are shown in Fig. 9.11 for cells distributed on a polytetrafluorethylene surface. The distribution of cells was calculated for an 8 μm^2 unit area, which is close to the square of a single cell spread on friendly substrate. The probability of cell distribution for an untreated polytetrafluorethylene surface has a maximum corresponding to cell islands with a single cell square of 1 μm^2. With fluence of ion beam implantation, the peak of probability corresponding to islands disappears and the probability of cell distribution density becomes maximal at one cell per unit. This means that the cells are spread up to 8 μm^2 square for one cell.

An impressive spreading of the cells on a treated surface can be observed on materials made of fiber. For example, we observed cells on an expanded polytetrafluorethylene (ePTFE) surface. This material has the same chemical structure and properties as polytetrafluorethylene film, but

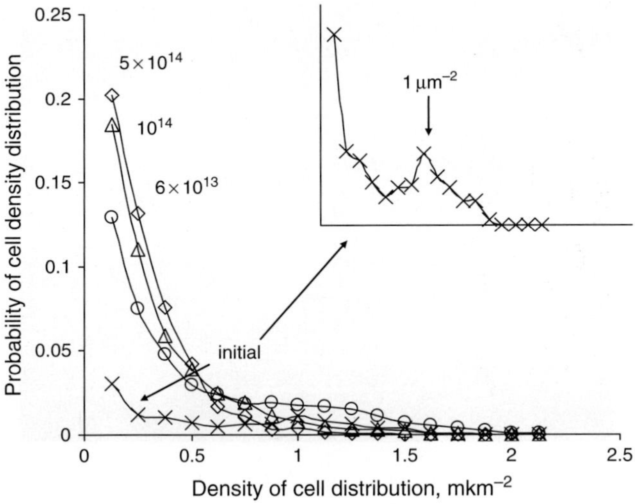

Fig. 9.11. Cell distribution on polytetrafluorethylene surface treated with ion beam: crest – untreated, circle – 6 × 10^{13} ions/cm^2, triangle – 10^{14} ions/cm^2, rhomb – 5 × 10^{14} ions/cm^2 dose of treatment. The experiments were done in cooperation with I. Kondyurina and M. Maitz.

it has a felt structure containing thin (μm) fibers. The cells on untreated ePTFE fibers sit compactly in local places to minimize the contact area with the fiber surface (Fig. 9.12). The cells on ion beam modified ePTFE fibers fill the whole surface of the fibers (Fig. 9.13).

The Alamar Blue assay was used to test for living cells on a polytetrafluorethylene surface. When Alamar Blue assay is added to solution with seeded cells, the assay penetrates into living cells but does not penetrate into dead cells. UV-vis spectra show that a concentration of assay remained in solution (Fig. 9.14). For example, the results of living cell analysis are presented for polyethylene after PIII treatment. For untreated (initial) polyethylene the number of living cells was minimal; and although the assay was not collected, its line at 630 nm was maximal. With fluence of PIII treatment of polyethylene, the amount of living cells increases and assay line intensity decreases. The absorbed value of the Alamar Blue assay calculated as the difference between

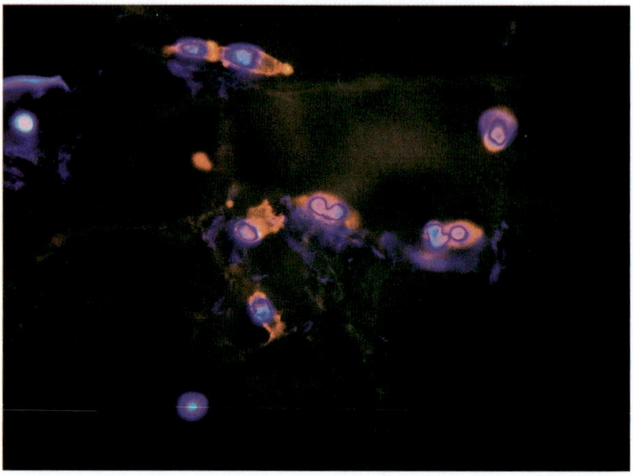

Fig. 9.12. Image of cells on untreated ePTFE substrate. Cells are compact. They cover part of ePTFE fibers. Experiments were done with M. Maitz by contract with Boston Scientific.

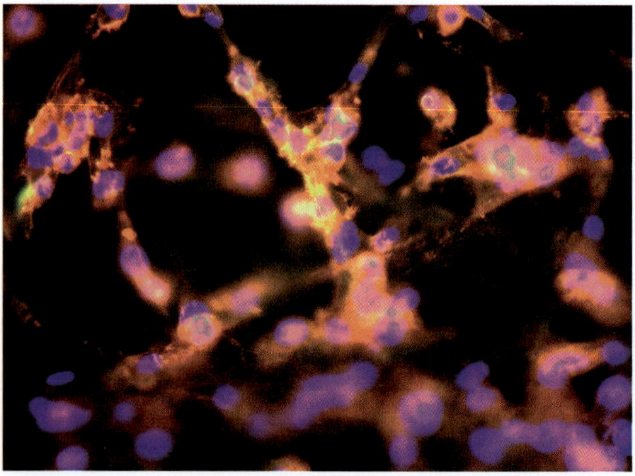

Fig. 9.13. Image of cells on ion beam modified ePTFE substrate. Cells are wide and spread on all fibers. Experiments were done with M. Maitz by contract with Boston Scientific.

buffer absorbance at 570 nm and Alamar Blue absorbance at 630 nm shows activity of the cells adhered to the polyethylene surface with fluence of PIII treatment (Fig. 9.15). Cell activity increases if the polyethylene substrate is modified,

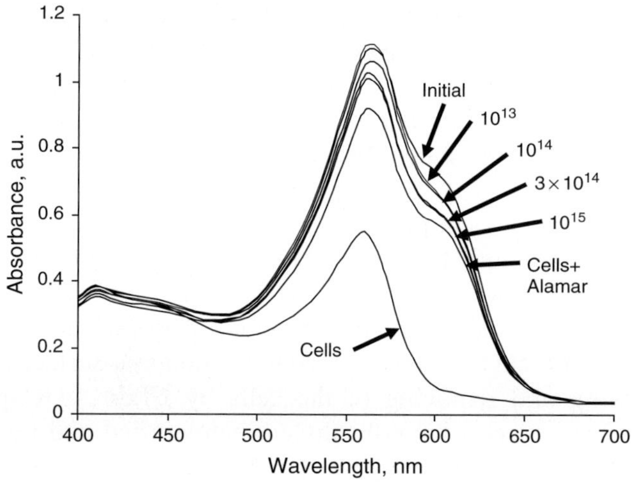

Fig. 9.14. Alamar Blue array spectra of cell media after growing on polyethylene substrate treated with PIII. Measurements were done after 6 hours of growing. The experiment was done in cooperation with I. Kondyurina and M. Maitz.

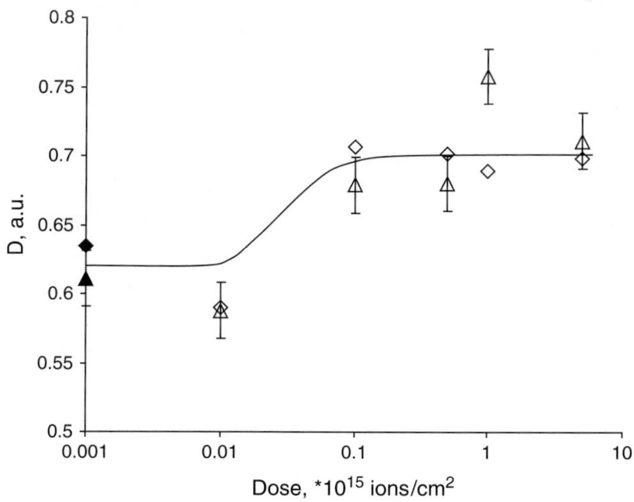

Fig. 9.15. Alamar Blue array absorbance in cell media after growing on polyethylene surface modified with PIII. Measurements were done after 6 hours (rhomb) and 8 hours (triangle) of growing. The experiment was done in cooperation with I. Kondyurina and M. Maitz.

but after barrier fluence of 10^{14} ions/cm^2 cell activity does not change. At this fluence, sufficient modification of the polymer surface occurs such that complete covering of the substrate with living cells is achieved – essentially the polymer becomes a friendly substrate for cell life and growth.

If the attachment of the cells occurred through the attachment of proteins, the proteins must be remain stuck to the polymer surface after removing the cells, and indeed for protein, this was observed in previous experiments. In reality, the proteins remain attached to the polymer surface after growing and removing of the cells by FTIR ATR spectra. Such spectra for polyethylene unmodified and treated by PIII are shown in Fig. 9.16. The spectrum of untreated polyethylene contains weak lines of proteins. The spectra of PIII modified samples contain strong lines of Amide A, Amide I and II proteins. Such spectral peaks do not disappear after washing in water and detergents. The proteins are stuck covalently as in previous experiments with protein solution.

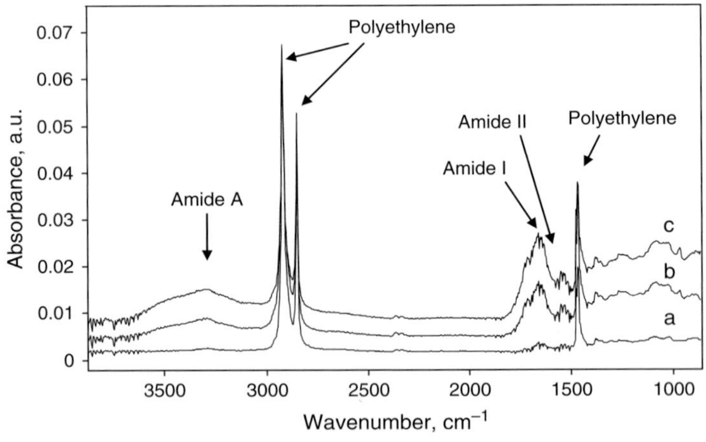

Fig. 9.16. FTIR ATR spectra of polyethylene after cell growing and detergent washing: a – untreated surface, b – treated with 10^{15} ions/cm^2 PIII dose, c – treated with 10^{16} ions/cm^2 PIII dose.

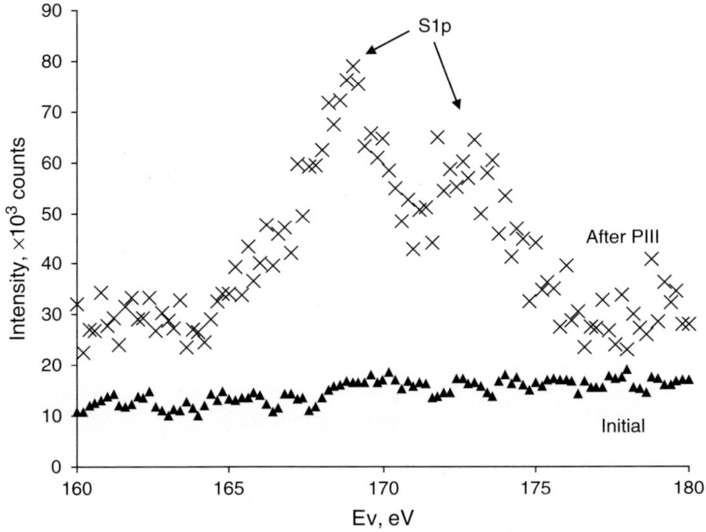

Fig. 9.17. XPS spectra of two PTFE surfaces after cell growing and detergent washing: untreated sample and sample modified with PIII.

Additionally, XPS spectra of polyethylene showed protein attachment to PIII treated surfaces: spectra of S2p line for polytetrafluorethylene are shown in Fig. 9.17. The spectrum of untreated polytetrafluorethylene does not contain the sulfur line after cell attachment. This means that the proteins are not detected for the untreated surface. The spectrum of the PIII treated surface shows peaks which indicate the presence of the sulfur atoms of the protein molecules and confirms that the protein is strongly attached to the modified polymer surface.

The same spectral lines appear in spectra of polyethylene after implantation into a living organism (Fig. 9.18). Polyethylene samples were inserted into a rat organism at surgery. The samples were placed under skin of wall. After six months the samples were removed at surgery and washed. The FTIR ATR spectra were recorded. The spectrum of the untreated sample did not contain any additional lines except polyethylene macromolecule lines. The spectrum of the ion beam treated sample contained Amide A,

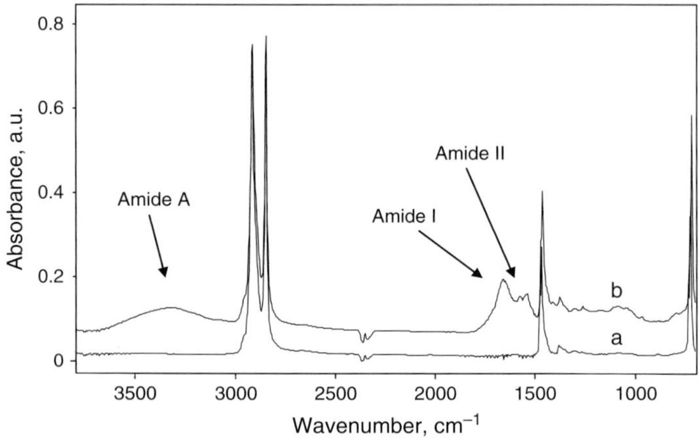

Fig. 9.18. FTIR ATR spectra of polyethylene which was implanted in rat organism during 6 months: a – untreated sample, b – treated with ion beam with 10^{15} ions/cm^2 dose.

Amide I and II lines, corresponding to vibrations of proteins attached on the modified surface. Such protein lines remained after deep washing of the samples with water, detergents and solvents such as ethanol and toluene. Therefore, the proteins were stuck covalently on the modified polyethylene surface as observed in cell and protein experiments.

Covalent bonding to the polymer surface, and retention of active protein conformation is very important. The modified surface after ion beam becomes attractive to any kind of proteins. In living organisms the modified polymer is covered by its own proteins immediately after first contact with the organism media. Covering is complete after protein attachment when the polymer surface does not contact directly with the organism media, including phagocytes, leukocytes, macrophages and other components of the immune system. All organism cells contact with proteins adsorbed on the polymer surface. This means that the organism's immune system cannot recognize the artificial polymer implant covered by its own proteins and does not isolate it from the organism media. The polymer implant becomes its "own" organ because it is covered by its "own" proteins.

Ion beam implantation was also applied to industrial polymer implants which can be used in humans. Using polymer implants as human organ replacement has become far more significant over the last few decades primarily due to: the development of new materials which have similar properties to human tissue; and the development of new methods in surgery such that a larger number of human organs can be replaced. One of the important fields of artificial organ implantation is mammary prosthetics.

In modern cosmetic surgery, three polymer materials are usually used for mammary prosthetics: silicon rubber, polyacrilamide gel and polyurethane foam. Polyurethane material is used less, but is being considered more and more. The problem with silicon rubber for human organisms is due to the difference between silicon and human tissue. Biological tissue naturally degenerates. Death of an organism's cells and destruction of the cell's molecular components are fundamental processes in a cell's lifecycle. The products of such degradations are recycled and used as building blocks for new proteins, sugars, lipids and other molecules as well as cells. The process of regeneration of human and animal organisms by themselves is part of life. However, silicon is a foreign substance and cannot be utilized by an organism's regeneration system. The destructive products of silicon cannot be released from an organism in the normal way through urinary organs; such products can only be collected elsewhere in the organism. Apart from inertness of silicon destruction products, their circulation and collection can have a negative influence on the functionality of the organism as a whole. Therefore, prolonged contact of organism media with silicon prosthetics is not preferable.

Another situation arises with polyacrylamide gel and polyurethanes. These polymers degrade with the formation of amines, amides and glycols, which can be utilized in biochemical reactions of organisms or released from organisms in the usual ways. The products of destruction are not collected in the organism. However, such products can cause toxic reactions, if the concentration of the products is higher than the toxic level. Therefore, the problem of biocompatibility for such polymers is connected with the rate of biodestruction.

Special medical polyurethane is used as a material for the shell of breast prosthetics, the shell of testicle prosthetics, the joint of metacarpophalangeal prosthetics and the elastic matrix of diaphragm prosthetics. Polyurethane based on polyoxypropylene/polyoxyethylene therminated with diisocyanate and cured with diamine is permitted for medical applications inside human organisms. Such polyurethane has a domain structure containing hard urea/urethane groups with strong hydrogen bonds and a polyether soft matrix providing elastic and strength properties for the prosthetics.

After polyurethane prosthetics are implanted, the organism initiates an immune response and surrounds the foreign body with a layer of collagen, in order to isolate itself from the foreign material. The collagen covering is stable and strongly attached to the polyurethane surface. Exposed polyurethane shells cannot be washed of collagen by any detergents without the chemical degradation of proteins. The problem is that the polyurethane has similar chemical properties and active chemical detergents destroy the polyurethane surface together with the protein covering. The formation of collagen covering may be detected with using the FTIR ATR. The spectra recorded from polyurethane samples after exposure to the human body contained collagen lines (Fig. 9.19): different Amide I lines were seen for polyurethane and protein. It was found that the collagen structure grew over a 5–6 months timeframe of exposure to the organism before stabilizing (Fig. 9.20). The same effect of collagen structure formation is observed at exposure in rat organisms (Fig. 9.21). The spectral lines of the collagen covering grow more quickly than for spectra of the polyurethane shell in human organisms. However, the character of spectra change is the same. The collagen structure

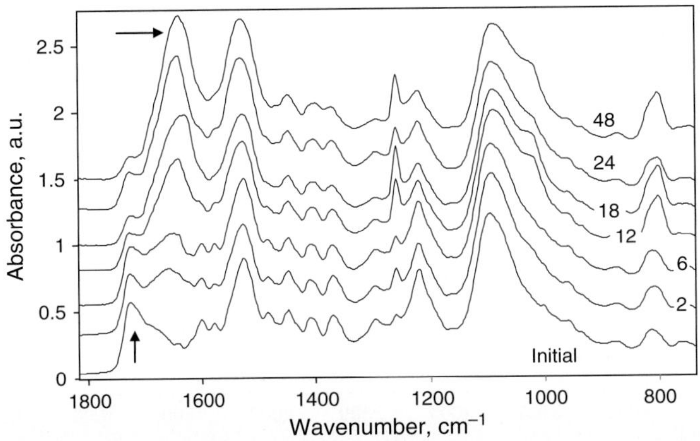

Fig. 9.19. FTIR ATR spectra of polyurethane shell of mammary prosthetics after ageing in human organism (months), arrays show Amide I peak positions for polyurethane and proteins. Experiments were done in cooperation with I. Osorgina and S. Plaksin.

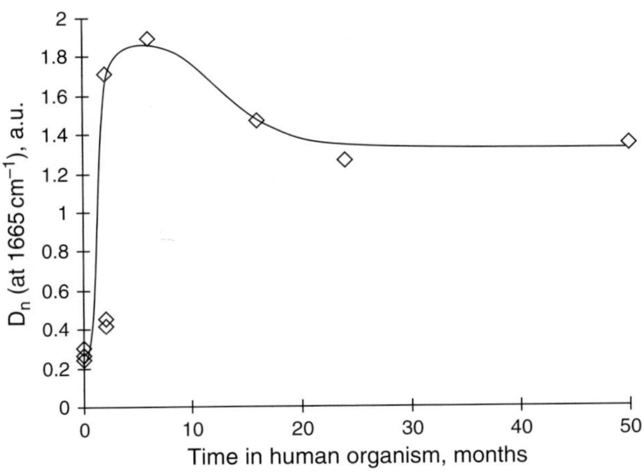

Fig. 9.20. Normalized absorbance of 1665 cm^{-1} Amide I line of collagen structure in FTIR ATR spectra of polyurethane shell of mammary prosthetics with time in human organism.

is viewed with an optical microscope as a network covering the polyurethane surface (Fig. 9.22).

After ion beam implantation the behavior of the polyurethane in the living organism changes significantly.

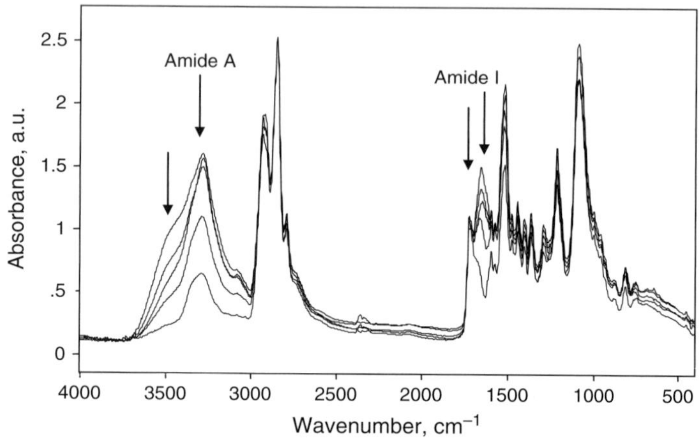

Fig. 9.21. FTIR ATR spectra of polyurethane shell of mammary prosthetics after ageing in rat organism. From bottom to top: initial polyurethane, after 10 days', 1 month's, 5 months' and 9 months' ageing in rat organism. Experiments were done in cooperation with I. Osorgina and I. Kondyurina.

FTIR ATR spectra show a significant decrease in the amount of collagen shell stuck to the polyurethane surface (Fig. 9.23). The amount of collagen was estimated by relative intensity of Amide I and Amide A lines in FTIR ATR spectra and decreased significantly in ion beam modified polyurethane samples (Fig. 9.24).

In addition to spectral results the absence of collagen shell on the ion beam modified surface is observed using optical microscopy (Fig. 9.22). The polyurethane surface modified with high fluence of ion beam implantation has a network of cracks, which appears as a result of the cracking of brittle carbonized layers of polyurethane after ion beam. After ageing in the organism, the cracks become deeper due to biodestruction processes in unmodified areas of polyurethane. However, the surface of the carbonized area is not changed and it does not contain collagen structure on the top. At low fluence of ion beam implantation, the network of cracks is very weak. Carbonization is not complete and the carbonized layer is not brittle enough for

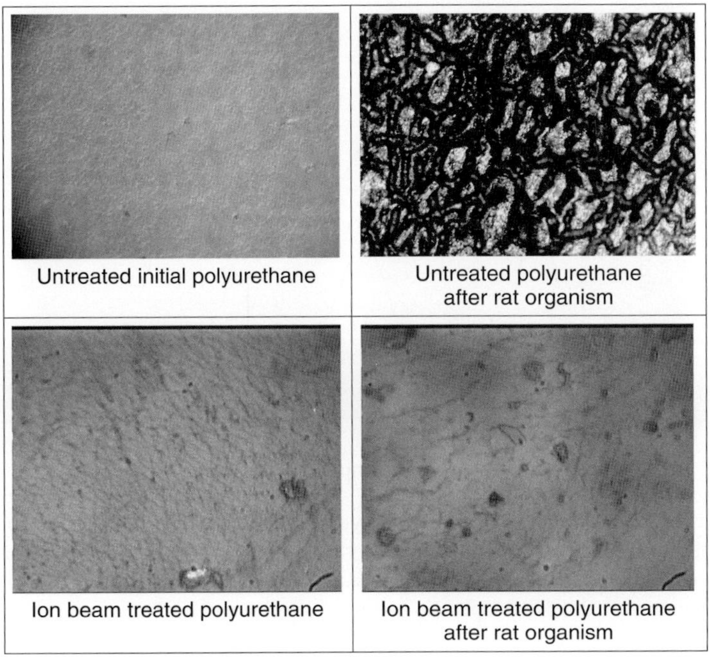

Fig. 9.22. Microphotographs of polyurethane shell of the prosthetics after ion beam implantation and exposure in rat organism during 5 months.

cracking. Also after ageing in the organism, polyurethane does not contain collagen shell on the top of surface and the cracks do not change.

This behavior of the polyurethane surface can also be observed using scanning electron microscopy (Fig. 9.25). The untreated surface of polyurethane contains collagen structures which cover the surface. After ion beam implantation the collagen is observed in some areas close to the cracks, where untreated polyurethane surface could be reached by leukocytes and macrophages. The top surface of the carbonized layer is free from collagen.

A simple schematic diagram of the polymer implantation process is shown in Fig. 9.26. Untreated polymer is recognized by the immune system of the organism as a foreign body and the organism builds a wall (shell) to isolate the

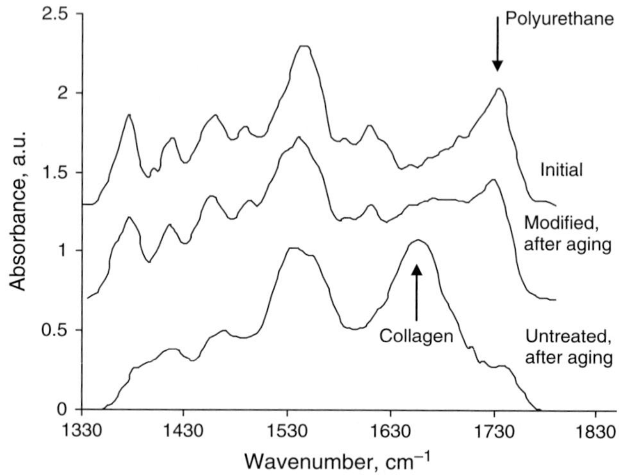

Fig. 9.23. FTIR ATR spectra of polyurethane shell after ion beam implantation and ageing 4 months in rat organism. Initial spectra recorded from fresh polyurethane shell. Dose of treatment is 5×10^{14} ions/cm^2. Experiments were done in cooperation with I. Osorgina and I. Kondyurina.

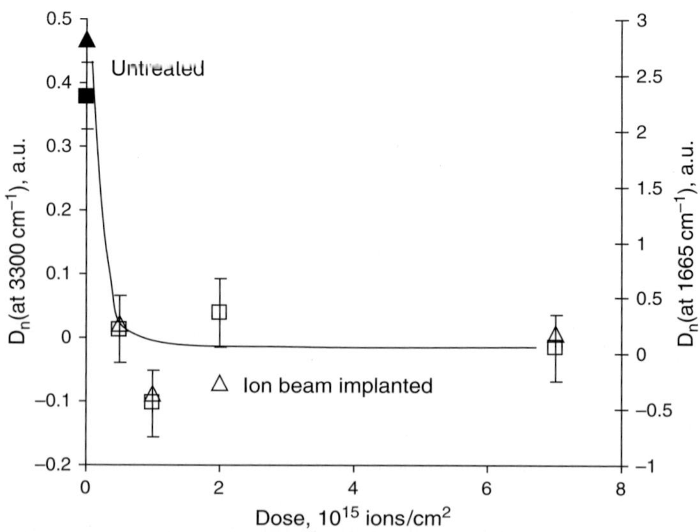

Fig. 9.24. Normalized absorbance of 3300 cm^{-1} Amide A line (triangle) and 1665 cm^{-1} Amide I line (cubic) of collagen structure in spectra of polyurethane shell after ion beam implantation and ageing in rat organism.

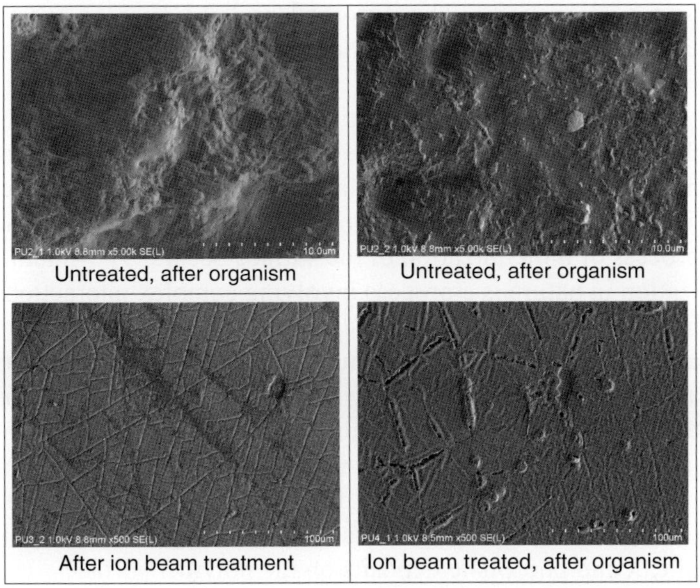

Fig. 9.25. Electron microscopy photographs of polyurethane surface implants after 5 months' ageing in rat organism. Experiments were done in cooperation with I. Osorgina and I. Kondyurina.

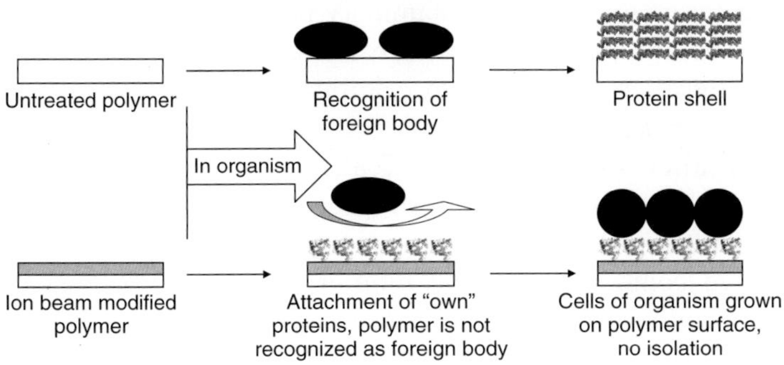

Fig. 9.26. Initial and modified polymer is installed into living body. Simplified scheme of organism reaction.

foreign body from the organism media. In the case of ion beam modified polymer, the proteins of the organism are adsorbed on the polymer surface, cover it completely and the organism immune system cannot recognize the polymer as a foreign body. The adsorbed proteins enables organism's cell attachment and the polymer becomes ingrown into organism's tissues.

The modified polymer can find specific applications in the human body as heart valves, artificial vessels, urea channels, mammary prosthesis, face and finger implants, etc. Together with protein absorption and cell attachment the process of mineralization of the surface occurs [9]. Mineralization in nature is a process involving complex interactions between inorganic ions, crystals and organic molecules in an aqueous media. The organic and inorganic counterparts are continuously interacting with each other and the result is the formation of exquisite mineralized tissues with convoluted architecture and remarkable, microscopic designs. For polymers used as soft tissue implants in living organisms, the biomineralization process – specifically the formation of calcium phosphate deposits on the polymer surface – can induce characteristic structures that alter the success of the polymer implant. For example, calcium phosphate deposits on the internal sides of artificial vessels can obstruct the blood circulation; calcification can decrease the elasticity of soft tissue implants like mammary prosthesis, which brings pain to the patient. Heart valve operations can be unsuccessful after a time due to calcification of the valve walls which make them non-elastic.

If after ion beam modification the polymer surface becomes active to adsorbed protein molecules, there is a danger that the activated polymer surface will accept active components of the biomineralization process, and more intensive calcification processes will lead to impossibility of modified polymer using. We have investigated the mineralization process after ion beam modification for polyethylene, polytetrafluorethylene, and some kinds of polyurethanes, which are used for medical applications.

The biomineralization process of polymers was performed with a simulated body fluid (SBF) technique, which simulates the human blood plasma. The modified polymers were immersed in a supersaturated solution for 4 hours at 37°C prepared just before samples were immersed. Two base solutions were prepared with reagent-grade chemicals. The first one consisted of NaCl (15.99 g/l), KCl (0.45 g/l), $CaCl_2 \cdot 2H_2O$ (0.74 g/l) and $MgCl_2 \cdot 6H_2O$ (0.61 g/l) dissolved in 1 l of distilled water. The second one included $Na_2SO_4 \cdot 10H_2O$ (0.32 g/l), $NaHCO_3$ (0.71 g/l) and $K_2HPO_4 \cdot 3H_2O$ (0.69 g/l) dissolved in 1 l of distilled water. The pH of the two solutions was buffered at 7.4 with tris-hydroxymethyl-aminomethane buffer. Before immersion of the samples, equal quantities of the two base solutions were mixed to obtain the final SBF solution. After immersion in SBF solution, the polymer samples were investigated by spectral methods and optical microscopy.

Microphotos of hydroxyapatite crystals on polyurethane surfaces are shown in Fig. 9.27. In addition to a dendrite structure, individual crystals were seen to cover the surfaces inhomogeneously. The crystals grown over a 4 hours duration had quite similar distribution density on the untreated and PIII modified polyurethane surfaces. The dendrite structure of the crystals supports that fact that the crystals grew from some local centers, which have low contentration on the polymer surface. When the growth of hydroxyapatite layer was initiated by the initial crystals, the subsequent hydroxyapatite crystal growth formed a structure that was mostly connected using the properties of the initial crystals

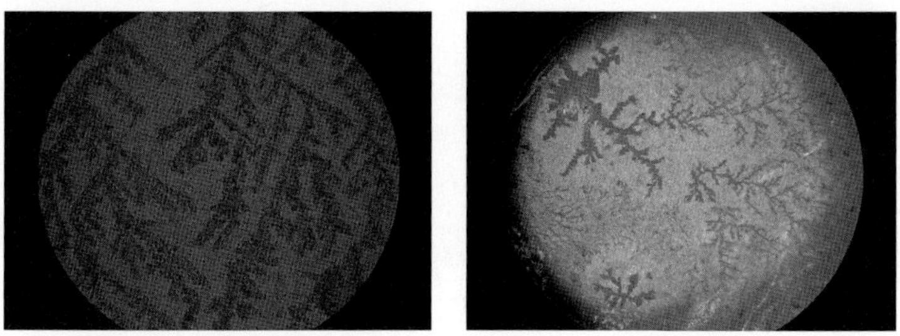

After PIII with fluence of 10^{16} ions/cm^2 — Untreated

Fig. 9.27. Microphoto of polyurethanes soaking in SBF solution: Untreated and PIII modified.

rather than those of the polymer surface. Despite the high concentration of morphological and structural defects on the polymer surface after PIII modification, significant differences in hydroxyapatite crystal growth (such as additional centers) on PIII modified polymer surfaces as compared to untreated polymer surfaces might be expected, however, no significant difference in hydroxyapatite crystal growth was observed. The same situation was observed for different kinds of polyurethanes based on polypropylene glycols and toluene diisocyanate, for commercial polyurethanes like Pellethane (Dow Chemicals), and for polytetrafluorethylene and polyethylene.

The structure of the crystals was analyzed by micro-Raman spectroscopy which gave vibrational spectra of the crystal thin layer on the top of the polymer surface. Focusing on the crystals was possible due to the micrometer size of the beam and allowed us to collect more signals from the layer with minimum influence of the underlying polymer. The recorded Raman spectra of the PTFE surface with hydroxyapatite crystals show additional lines at 428.5, 561.2 and 954.4 cm^{-1} present in the top spectrum

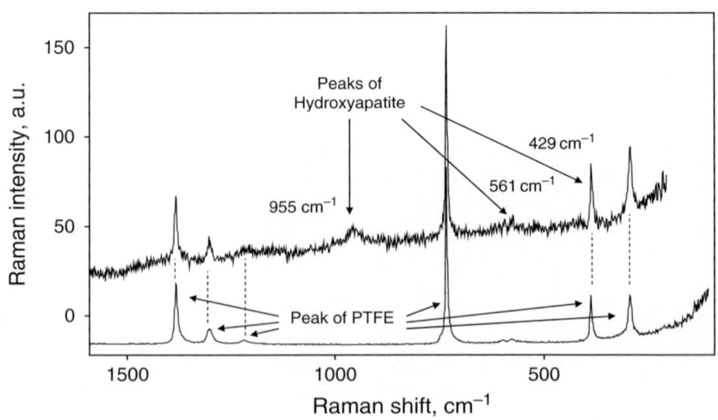

Fig. 9.28. Micro-Raman spectra of PTFE surface: bottom is initial PTFE; top is PTFE surface layer after PIII and soaking in HA solution. Experiment was done in cooperation with E. Pecheva.

(Fig. 9.28). The peak at 428.5 cm^{-1} is overlapped with the peak of PTFE at the same position. The peaks are assigned to ν_2 asymmetric bending, ν_4 symmetric bending and ν_1 symmetric stretching vibrations of PO_4 in calcium phosphate. Generally, ν_1 stretching mode is the strongest peak in the spectrum and it is very narrow, thus indicating good crystallinity. The broad peak in our spectrum was ascribed to the lower degree of crystallinity of our layer which has not matured. It is known that at precipitation from aqueous solution, before the final formation of crystalline hydroxyapatite, other precursor phases appear. These are amorphous calcium phosphate, octacalcium phosphate and/or dicalcium phosphate dehydrate, and low crystalline hydroxyapatite.

The spectral lines of the hydroxyapatite structure are observed in FTIR spectra of polymers after the PIII modification and immersion in the SBF for 4 hours (Fig. 9.29). After immersion of the polymers in the SBF, a weak peak at 1050 cm^{-1} due to the contribution of the ν_3 P—O asymmetric stretching mode of calcium phosphates (PO_4) was found in the spectra. This shows that the calcium phosphate layer was grown on the control and PIII treated LDPE surface from the supersaturated aqueous solution. The relative

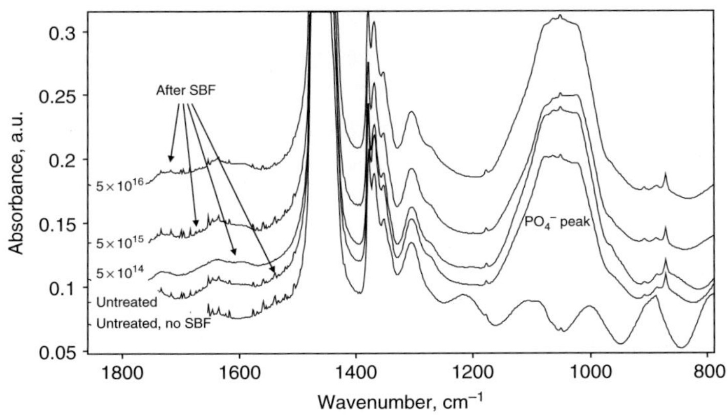

Fig. 9.29. FTIR transmission spectra of 50 μm polyethylene film after PIII and soaking in SBF. PIII dose is noted in ions/cm^2. Experiment was done in cooperation with E. Pecheva.

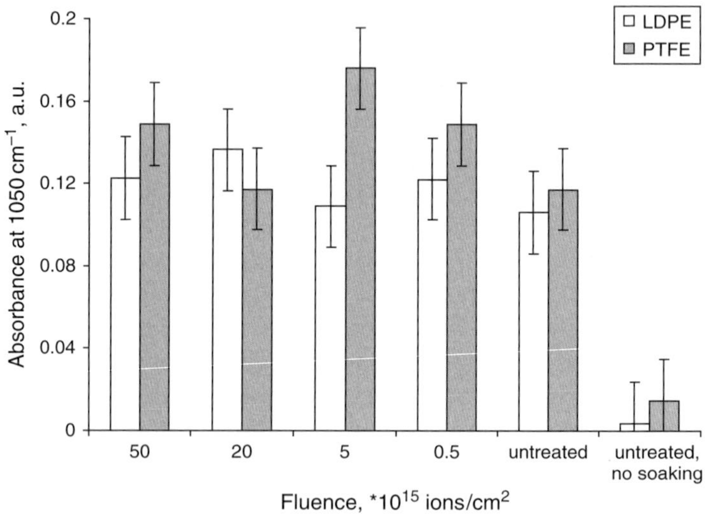

Fig. 9.30. Absorbance of hydroxyapatite peak at 1050 cm^{-1} in FTIR transmittance spectra of LDPE and PTFE films after PIII and soaking in SBF. Experiment was done in cooperation with E. Pecheva.

thickness of hydroxyapatite layer thickness could be estimated based on the intensity of this peak in FTIR spectra. Using Beer's law the absorbance at 1050 cm^{-1} is interpreted as proportional in value to the hydroxyapatite thickness (Fig. 9.30). The thickness of the hydroxyapatite on modified and initial surfaces is the same in a frame of error bars. The amount of precipitated hydroxyapatite on polyethylene and polytetrafluorethylene surfaces does not depend on fluence of PIII treatment.

The measurements of the hydroxyapatite precipitation were done after a short time of polymer soaking in SBF. At such stage of precipitation, the amount of hydroxyapatite crystals depended on centers of precipitation placed on the polymer surface. Subsequent growth of hydroxyapatite crystals does not depend on surface properties. The equality of spectral and optical data for the amount of hydroxyapatite crystals on unmodified and modified polymer surfaces show that ion beam implantation does not affect significantly initiation of hydroxyapatite crystal growth, which is the main reason for calcification of polymer implants in organisms. Therefore, ion beam implantation of polymer implants does not increase the risk of calcification.

Ion beam implantation has been found useful for very specific applications in medical polymers. For example, ion beam implantation was applied for regulation of drug release from polymer coating of vascular stents. The application of vascular stents in combination with angioplasty of arteriosclerotic blood vessels has become a standard technique, which is widely equivalent to bypass surgery. However, the application of a stent in the body often induces a local inflammatory reaction, proliferation of the smooth muscle cells and neointima formation. This reaction can be reduced by application of a stent with a polymer coating which contains anti-inflammatory or antiproliferative drugs. These drugs are released over a long time and get effective concentrations only locally at the vessel wall without achieving effective concentrations systemically.

If polyurethane were soluble, the drug could be incorporated in polyurethanes through solution. The drug is added to a solution of the polyurethane. After evaporation of the solvent a layer of polyurethane incorporating the drug is formed. If the drug is inactive in a reaction of polyurethane synthesis, the drug could be inserted into a reactionable mixture of the components before the synthesis of polyurethane. After reaction, the polyurethane contains the drug uniformly distributed throughout the polyurethane coating.

If polyurethane is crosslinked and cannot be dissolved, the drug could be inserted by swelling of the polyurethane in a drug solution. After evaporation of solvent the drug remains in polyurethane network.

The kinetics of drug release from a polymer to blood is described generally by the so-called "frontal process" of water diffusion into the polymer. According to this model the release process is determined by the water front, which penetrates the polyurethane when exposed to water. The rate of the frontal process depends on the water diffusion coefficient into the polyurethane, on the maximal swelling of polyurethane in water and on the solubility of prednisolone in water.

The first theoretical model of the drug release process was developed by T. Higuchi [10]. By experimental estimation, 80% of the drug is released by the frontal law and 20% of the drug amount is released by diffusion [11]. In general the kinetics of the drug release from a swelling polymer is very complex. It depends on the solubility and the swelling of the polymer in water, the concentration limit of the drug in water and the distribution of the drug in the polymer [12]. For thin coating, it gives inhomogeneous kinetics of drug release, when the drug concentration at first period of release is high and then decreases sharply. Sometimes, the drug release is so quick that the concentration of the drug becomes lower than the therapeutical effect level. However, the drugs usually have a toxic level of concentration in the organism. In such a case, the concentration of the drug in polymer cannot be increased high enough to exclude the toxic level of drug in the organism. The required drug concentration therefore lies in a region, with upper and lower limits determined by the toxic level and therapeutic effect level, respectively. The kinetics of drug release must be as constant as possible.

We used ion beam implantation of polymer coating to achieve a more uniform drug release kinetic. For example, we used PIII modification of polyurethane to change the kinetics of drug released. The polyurethane was synthesized from a co-polymer of polyoxytetramethylene/polyoxypropylene glycol terminated by 2,4-toluene diisocyanate and crosslinked with 3,3′-dichlor,4,4′-diaminediphenylmethane. After synthesis, the crosslinked polyurethane was swollen up to saturation stage in a solution of ethanol with prednisolone as drug agent. After swelling, the polyurethane was dried: ethanol left prednisolone in the polyurethane and evaporated. The prednisolone release kinetics from polyurethane to water solution was measured using the FTIR ATR spectroscopy method.

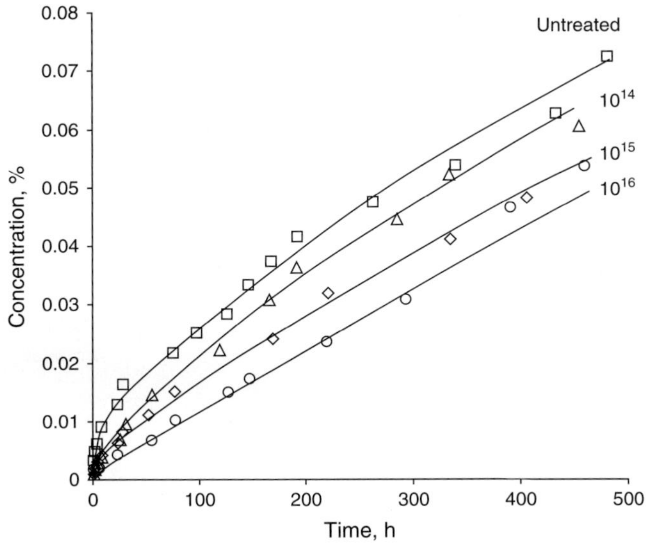

Fig. 9.31. Concentration of prednisolone in solution released from polyurethane coating which was modified with PIII. Fluence of PIII is noted in ions/cm^2. Experiments were done in cooperation with M. Maitz and I. Kondyurina.

The kinetics of prednisolone release from untreated polyurethane is not linear (Fig. 9.31). The kinetic curve of the prednisolone release corresponds to the frontal process of drug release from a solid matrix by the theoretical model of Higuchi [10]. The kinetics curve has a high slope at first period of the prednisolone release, then the slope of the curve becomes lower. According to the frontal process theory, the non-uniform release of prednisolone into water is observed due to a quick diffusion of prednisolone from the surface layer of the polymer. The diffusion from bulk layers takes longer and the rate of drug release becomes lower with time.

After PIII modification the kinetic curve becomes smoother. At high fluence of treatment (10^{16} ions/cm^2) of polymer the kinetics curve becomes close to linear without a sharp jump at the first period of release.

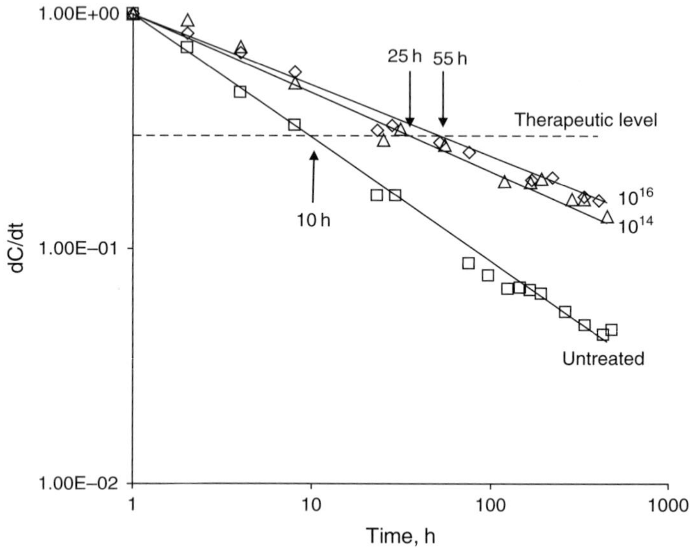

Fig. 9.32. Normalized rate of prednisolone release from polyurethane modified with PIII. Fluence of PIII is noted in ions/cm^2. Experiments were done in cooperation with M. Maitz and I. Kondyurina.

The rate of drug release from untreated polyurethane can decrease 40-fold after two weeks (Fig. 9.32). Therefore, the concentration of drug in blood vessels in place of the coated stent decreases by the same value. Prednisolone has a sufficiently narrow diapason of concentrations between toxic and therapeutic levels. If we take into account the difference between toxic and therapeutic concentrations, the period of active prednisolone concentration after inserting of untreated polymer implant is about 10 hours. For a PIII modified polyurethane implant, the period of active prednisolone concentration is 25 h and 55 h for 10^{14} and 10^{16} ions/cm^2 fluence of PIII, respectively. The prolongation of the active effect of prednisolone decrease the risk of inflammatory reaction, proliferation of the smooth muscle cells and neointima formation caused by inserting the vascular stent.

References

1. K.B. Lee, S.J. Park, C.A. Mirkin, J.C. Smith, M. Mrksich, Protein nanoarrays generated by dip-pen nanolithography, *Science* 295 (2002), 1702–1705.
2. N.J. Nosworthy, J.P.Y. Ho, A. Kondyurin, D.R. McKenzie, M.M.M. Bilek, The attachment of catalase and poly-L-lysine to plasma immersion ion implantation-treated polyethylene, *Acta Biomaterialia*, accepted for publication.
3. B.K. Gan, A. Kondyurin, M.M.M. Bilek, Comparison of protein surface attachment on untreated and plasma immersion ion implantation treated polystyrene: protein islands and carpet, *Langmuir* 23 (2007), 2741–2746.
4. A. Kondyurin, N.J. Nosworthy, M.M.M. Bilek, Attachment of horse radish peroxidase protein to polytetrafluorethylene (Teflon) after plasma immersion ion implantation, prepared for publication.
5. M. Gajhede, D.J. Schuller, A. Henriksen, A.T. Smith, T.L. Poulos, *Nat. Struct. Biol.* 4 (1997), 1032–1038.
6. X. Su, Y. Zong, R. Richter, W. Knoll, *J. Coll. Interf. Sci.* 287 (2005), 35–42.
7. S. Krimm, Vibrational analysis of conformation in peptides, polypeptides and proteins, *Biopolymers* 22 (1983), 217–225.
8. G. Cohen, M. Kim, V. Ogwu, A modified catalase assay suitable for a plate reader and for the analysis of brain cell cultures, *J. Neurosci. Meth.* 67 (1996), 53–56.
9. L. Pramatarova, E. Pecheva, Modified inorganic surfaces as a model for hydroxyapatite growth, in *Materials Science Foundations*, vol. 26, Trans Tech Publications, Switzerland, 2006, pp. 1–122.
10. T. Higuchi, *J. Pharmac. Sci.* 52 (1963), 1145.
11. B.C. Anderson, S.K. Mallapragada, in American Physical Society, Meeting at March 2000, London, 2000.
12. D.S. Cohen and T. Erneux, *SIAM J. Appl. Math.* 58 (1998), 1193.

10 Protection in an aggressive environment

All polymer materials continuously destruct under environmental action [1–6]. The basic factors causing destruction under normal environmental conditions are sunlight, water (moisture) and air oxygen. In some applications polymer materials work under specific aggressive media such as solvents, chemically active gases or liquids, high energy irradiations such as UV and VUV light, electron and ion flux, and X- and γ-rays. The stabilization of polymer materials against degradation is very important area in the polymer industry.

Ion beam implantation changes the structure of the polymer surface layer. If the surface layer is responsible for the degradation process in aggressive environments, ion beam implantation could increase the resistance of modified polymer against aggressive environmental factors. Despite the high effectiveness of ion beam treatment for polymers we know only a few applications of ion beam implantation for polymer resistance improvement.

One of the important applications of stabilized polymers is the greenhouse, where a polymer film is exposed to sunlight (including low and high energetic UV light), oxygen and ozone active gases, gaseous industrial and natural pollutions of acids and bases, water and snow. Such outdoor ageing is very intensive in high sunlight irradiation regions such as deserts [4]. The problem of polymer stability in aggressive environments is usually solved by the application of special additives of antioxidants, fillers and inhibitors of ageing. But antioxidant and inhibitor activity is limited and sometimes lifetime is reduced under strong aggressive environments like intensive short-wave UV light in Australian deserts. In such cases other methods of preventing degradation processes become necessary.

Polyolefins are sufficiently sensitive to UV irradiations from the sun (Fig. 10.1). Under sunlight the polymer loses strength, becomes brittle and for some kinds of polyolefin the depolymerization process occurs. A mechanism of ageing under UV light includes an excitation of macromolecules at photon absorbance and a breaking of chemical bonds with the formation of free radicals. Such radicals start a chain of free radical reactions, which cause crosslinking, depolymerization and molecular structure transformation. In the presence of air, oxygen molecules take part in free radical reactions with the formation of oxygen-containing groups. The oxidation processes accelerate the ageing and depolymerization of polymers because of the free radical reactions with oxygen participation occurring quicker and causing new breaking of macromolecules. In the presence of oxygen without UV light polyolefins are sufficiently stable.

The modern antioxidants absorb the free radicals which are generated under UV light, and the stability of polyolefin

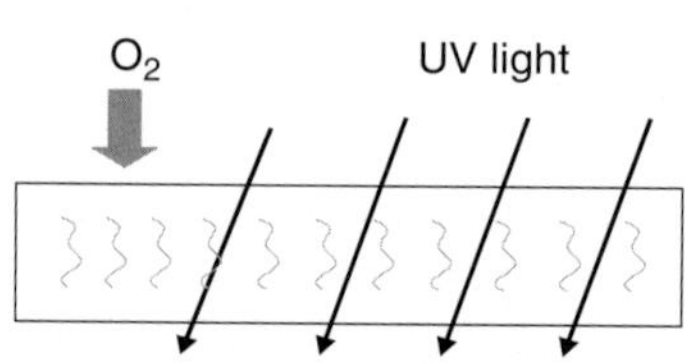

Fig. 10.1a. Ageing of untreated polyethylene in outdoor environment.

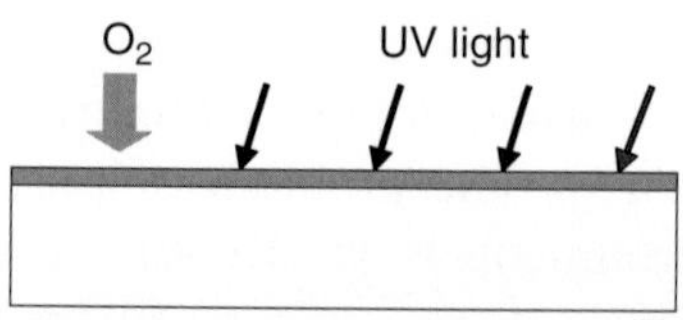

Fig. 10.1b. Ageing of modified polyethylene in outdoor environment.

is increased. There are usually special molecular traps for free radicals which stop free radical reactions at the start. Another method is a coating on important parts of polymer devices exposed to sunlight. But sometimes the coatings cannot be applied because the exploitation properties can be lost in the presence of additional coatings. In such special cases ion beam treatment can help.

In our experiments the low density polyethylene films of 50 mm containing N-phenylnaphthalene-1 as antioxidant were treated with nitrogen ion beam with 20 keV energy of ions. This film is usually used for greenhouse covers. The stability of the cover is about one year in an outdoor environment. After ion beam treatment the polyethylene films were exposed under real environmental conditions including sunlight, day and night variations in temperature, rain and snow, and moisture variations (Fig. 10.2). The films were exposed to open air in Mendeleevo village, Perm region, Russia. The distance from industrial centers was about 100 km; the town does not have any industrial pollution and car pollution is minimal. The samples were exposed for six months from July to January; average sun irradiation level was 85 kkal/cm^2, average annual temperature was +2°C, seasonal temperature variations were from −40°C to +30°C, average annual precipitation level was 450–500 mm including 60–80 mm of snow [7].

The ageing effect of untreated polyethylene is observed in FTIR spectra with the appearance of a carbonyl ν(C=O) band as the result of oxidation of polyethylene macromolecules under sunlight and air oxygen. The band

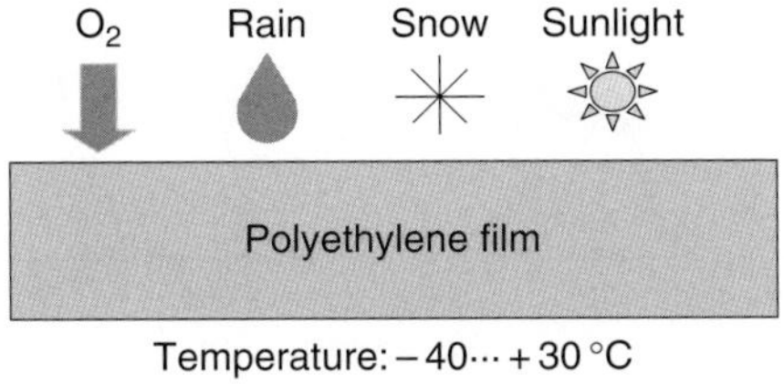

Fig. 10.2. Environmental test in outdoor environment for polyethylene film.

is wide and corresponds to overlapped individual carbonyl, carboxyl and aldehyde group lines. These groups appear in polyethylene macromolecules as a result of ageing and destruction. The intensity of such lines is attributed to the concentration of oxygen-containing groups which increase with an increase in the degree of ageing.

The wide band can be fitted with individual lines attributed to different chemical groups. The fitting is based on well-known literature data on carbonyl group spectra and a wide number of polymer oxidation investigations. FTIR spectra of aged polyethylene contains the 1708 cm^{-1} carboxyl group line, 1724 cm^{-1} ketone group line, 1740 cm^{-1} aldehyde group line and 1750–1770 cm^{-1} ester and γ-lactone group lines (Fig. 10.3). This corresponds to well-known spectra of polyethylene aged under UV light [1, 6, 8]. The band of polyethylene after ion beam is quite similar, because the processes of oxidation occur after ion penetration too (Fig. 10.4). But the intensity distribution and corresponding concentrations of oxygen-containing groups are quite different. The spectra of polyethylene after ageing contain more intensive low frequency shoulder attributed

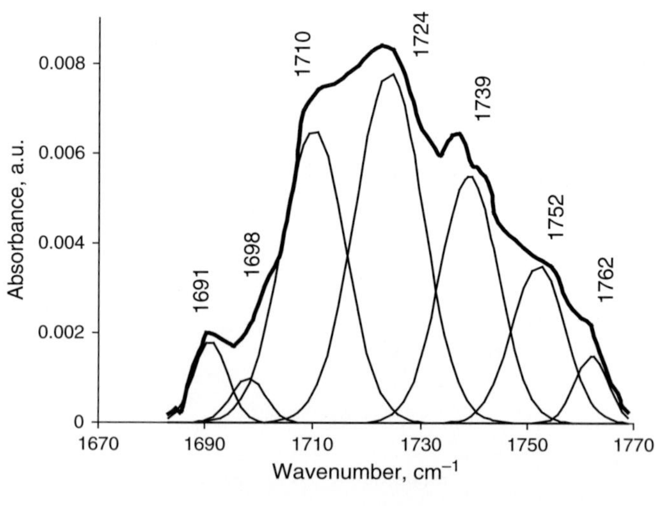

Fig. 10.3. Carbonyl band and its components in FTIR spectra of polyethylene after ageing in outdoor environment.

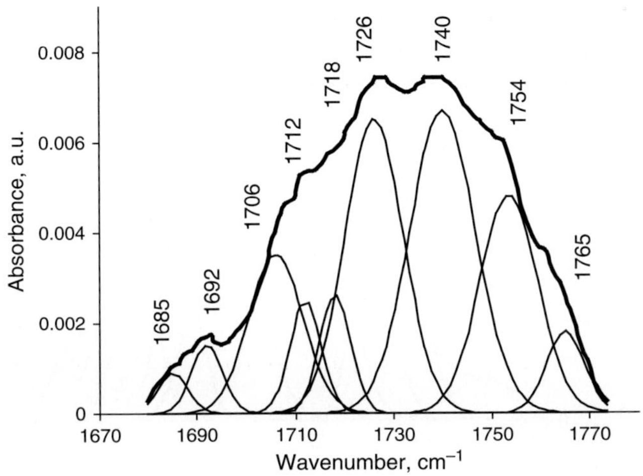

Fig. 10.4. Carbonyl band and its components in FTIR spectra of polyethylene after ion beam implantation.

to carboxyl and ketone groups. Polyethylene spectra after ion beam contain more intensive high frequency shoulder attributed to aldehyde and ester groups. The spectra difference is explained not only by the different oxidation processes, but also by the difference in structure of the surface layer of polyethylene. The surface layer of aged polyethylene contains only oxidized macromolecules which retain the structures of the polyethylene macromolecule. The surface layer of ion beam treated polyethylene contains highly carbonized structures. In such structures the oxidation process goes on another way and the oxygen-containing groups have different neighboring groups. After ageing of modified polyethylene the carbonyl band in FTIR spectra appears closer to the spectra of untreated aged polyethylene (Fig. 10.5). The high frequency shoulder decreases and the low frequency shoulder increases. This corresponds to the oxidation process in the unmodified bulk region of the polyethylene film.

The intensity of the carbonyl band increases with fluence of ion beam treatment as shown in Chapter 4, but

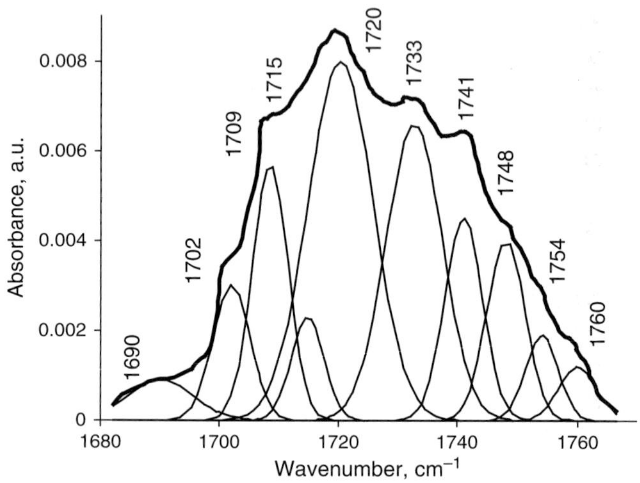

Fig. 10.5. Carbonyl band and its components in FTIR spectra of polyethylene after ion beam implantation and ageing in outdoor environment.

the intensity of the carbonyl line for ion beam modified polyethylene is lower than for aged polyethylene. The difference between oxidation of ion beam implanted polyethylene and polyethylene aged under outdoor environment gives an oxidation index. The oxidation index shows the amount of oxygen adsorbed and jointed to polyethylene macromolecules during ageing.

The oxidation index of polyethylene film aged six months is shown in Fig. 10.6 depending on ion beam implantation fluence. The normalized oxidation index determined by the carbonyl group line decreases with fluence of ion beam treatment. For polyethylene after ion beam implantation with 10^{14} ions/cm^2 fluence, the rate of oxidation determined by the oxidation index is 6 times lower than for untreated polyethylene film.

The strength of the polyethylene films decreases at ion beam treatment from 14.4 MPa for initial films to 10.8 MPa for 10^{14} ions/cm^2 fluence of treatment. This is caused by change of character of the stress/strain curve. More

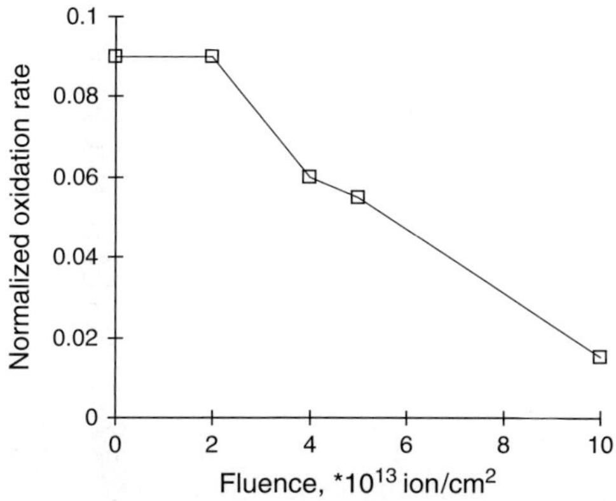

Fig. 10.6. Oxidation rate of polyethylene film after ageing at six months in outdoor environment.

detailed analysis of the strength changes is presented in Chapter 4.

After six months' ageing in an outdoor environment the strength of polyethylene film decreases from 14.4 MPa to 7.9 MPa. The polyethylene film becomes brittle and elongation at breaking decreases from 400% for initial films to 63% for aged films after six months in an outdoor environment.

The modified polyethylene film has different behaviors at ageing. The strength of the modified film decreases from 10.8 MPa to 10.2 MPa after ageing during six months. The elongation at breaking does not change after ageing. The strength degradation rate decreases with fluence of ion beam treatment (Fig. 10.7). At fluence of 10^{14} ions/cm^2 the mechanical properties of the polyethylene film remain the same despite exposure of the film to the outdoor environment.

The effect of ageing prevention is explained by absorption of UV light in the thin surface carbonized layer of modified polyethylene (Fig. 10.1b). The strong UV absorption is observed even at low fluence of treatment (see Chapter 4). On the other side, destruction is started from the surface

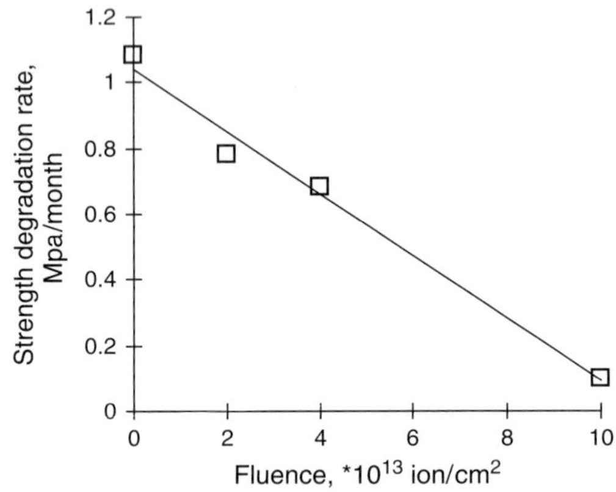

Fig. 10.7. Degradation rate of the polyethylene film strength at six months' ageing under outdoor conditions.

layer. In the case of modified polyethylene film the surface layer is carbonized and the carbon structure is more stable under destructive environmental conditions. Therefore the destruction process from the surface layer of modified polymer cannot propagate into the bulk layers. The chain reactions of free radicals do not propagate into the bulk layers and do not cause destruction of the whole polymer film. The modified layer works as a barrier to the destruction process in the bulk polymer layer. Therefore, the ion beam treatment causing the carbonization of the surface layer of polyethylene prevents the degradation of polymer film against outdoor environment factors.

Losing polymer exploitation properties can occur in aggressive liquid environments. For example, deformation of polymer is observed in aggressive (for polymer) solvent. Usually, a polymer material can be dissolved in a number of organic solvents. The solubility of polymer depends on intermolecular interactions between molecules of solvent and macromolecules of polymer. One of the characteristics of solubility is the Flory–Huggins parameter for solvents and polymers [9]. The Flory–Huggins parameter characterizes cohesion energy of a substance and similar Flory–Huggins parameters for solvents and polymers are a desirable condition for good solubility of polymer in solvent.

If a polymer does not have enough closed Flory–Huggins parameter to solvent parameter the swelling of polymer in solvent can be observed. This causes a deformation of polymer. Deformation is critical for polymer coating, when the deformation stresses can peel off the polymer coating from the substrate.

One of the possible solutions for excluding dissolving and swelling effects is a crosslinking of the polymer which has to be exposed to aggressive solvents. For example, this effect is used for rubber materials with plasticizers. In the absence of vulcanizing agents the rubber materials could even be dissolved in certain plasticizers. But vulcanized (crosslinked) rubbers have good mechanical properties and operate in the presence of active liquids such as lubricants or oil.

In specific cases vulcanization cannot be applied to polymer materials and methods of ion beam implantation could be used for crosslinking the polymer coating. We demonstrated this on an example of thin polystyrene coating. The polystyrene coating of 100 nm thickness was spin-coated on silicon wafer. Toluene is an excellent solvent for polystyrene and the toluene application takes only a few moments to completely remove the polystyrene coating from substrate. After ion beam implantation with low fluence the coating becomes crosslinked. Applied toluene could not remove any part of the polystyrene from the silicone wafer; the film is swelled and after drying the film is wrinkled due to evaporation of toluene (Fig. 10.8). The density of crosslinking is low and polystyrene macromolecules have enough mobility to accept toluene. After high fluence implantation, the polystyrene film does not change in toluene. The mass, thickness and surface morphology of the film are saved as initial. The ion beam implantation crosslinks the polystyrene macromolecules very densely which is enough to prevent toluene from swelling. Such a treatment can be used for specific coatings if other methods

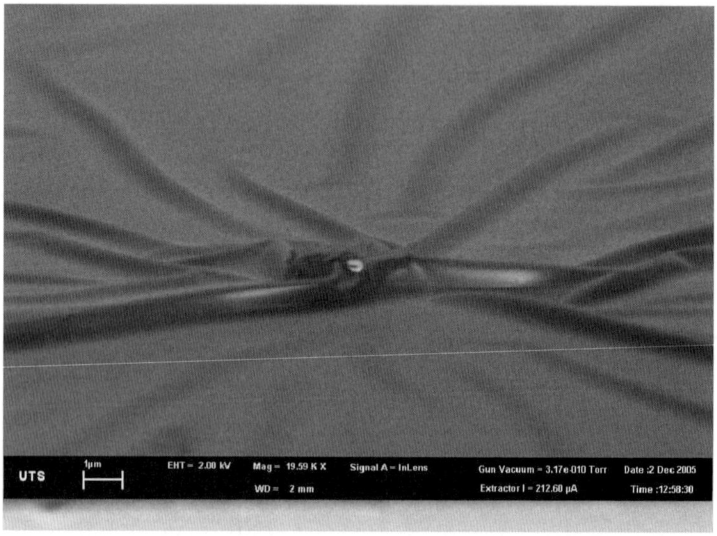

Fig. 10.8. Scanning electron microphotograph of polystyrene coating after PIII with low ion fluence and after swelling in toluene. SEM image was taken by R. Wuhrer. The experiment was done in cooperation with B.K. Gan.

of vulcanization or crosslinking cannot be applied or these are more complicated and expensive.

Ion beam modification can also be useful for polymers in chemically aggressive media. The modification of the thin surface layer cannot prevent complete degradation of the polymer, but some negative effects following the degradation process can be excluded.

An example is biodegradable polymers, which are used for drug delivery systems. The polymer contains a drug as a filler or in encapsulated volume. With time the polymer is degraded under hydrolysis reaction with water in organism media and the drug becomes free and releases to the organism media activating pharmaceutical effects in the organism.

Here we consider the degradation process of poly(lactic-co-glycolic acid) (PLGA) biodegradable polymer with many applications in medicine [10–15]. PLGA degrades in living organisms through a hydrolysis reaction in which degradation products, such as lactic and glycolic acids, are released:

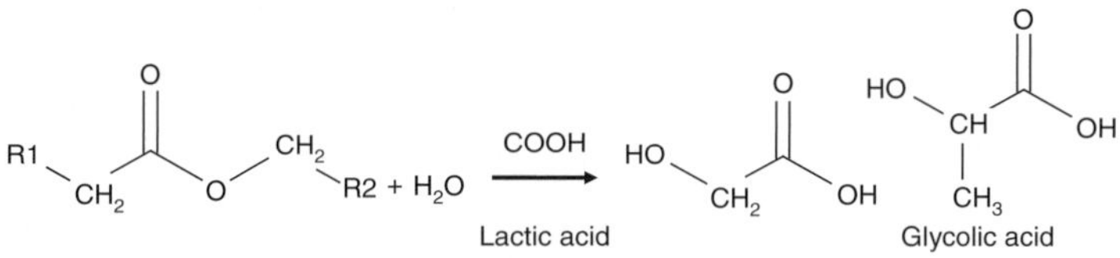

These reaction products are metabolized by the organism. Due to its safe biodegradability, the drugs encapsulated in the PLGA are gradually and locally released during the biodegradation process. PLGA can be used as a therapeutic agent in humans, in the form of drug delivery systems, tissue engineering implants, microspheres, microparticles, nanoparticles and coatings.

But PLGA is swelled in water solution before degradation. The swelling increases a dimension of the PLGA film. If PLGA coating has insufficient adhesion to substrate, the PLGA film ruptures and forms separate drops on the substrate surface (Fig. 10.9). The process occurs over a number of days. After some hours in water the surface develops hills and valleys. The valleys become deeper with time and eventually reach the substrate surface. The valleys spread until they overlap and separate drops of PLGA appear. This process is called "dewetting". Dewetting depends on interaction with substrate, viscosity of the film and presence of structure defects (rupture centers). The dewetting process is quicker than the degradation process. After dewetting the PLGA drops degrade by the way of hydrolysis. If dewetting of PLGA film occurs in an organism, the kinetics of drug release becomes uncontrolled, and the therapeutical effect of the drug becomes unpredictable.

We used PIII modification of PLGA films for excluding the dewetting effect of the PLGA coating in water solution. The surface morphology of the film does not change after ion beam treatment: the surface roughness remains the same.

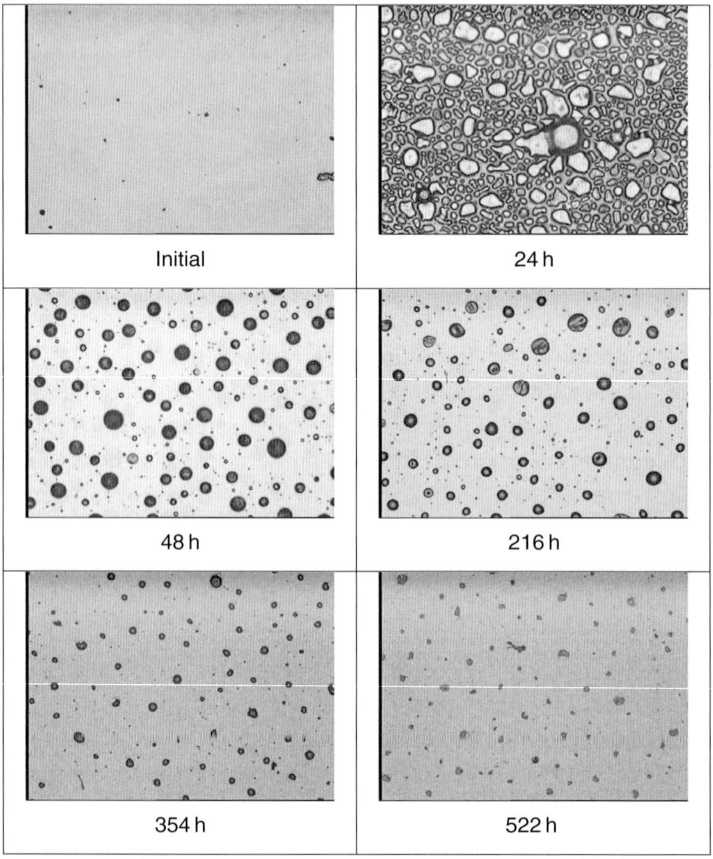

Fig. 10.9. Microphotos of untreated PLGA coating on silicon substrate with time of degradation in water. Size of photos is $1200 \times 950\,\mu m$. Experiments were done by I. Kondyurina.

Visual inspection of modified PLGA showed color changes: PLGA films subjected to high fluence of PIII become milky white when observed in reflection and light brown in transmission. The structure of PLGA film is changed under ion beam, which is observed by FTIR spectra of treated PLGA coating (Fig. 10.10). The absorbance in the 1600–1700 cm^{-1} region of the spectra increases with fluence of ion beam implantation. Using ellipsometrical data the refractive index

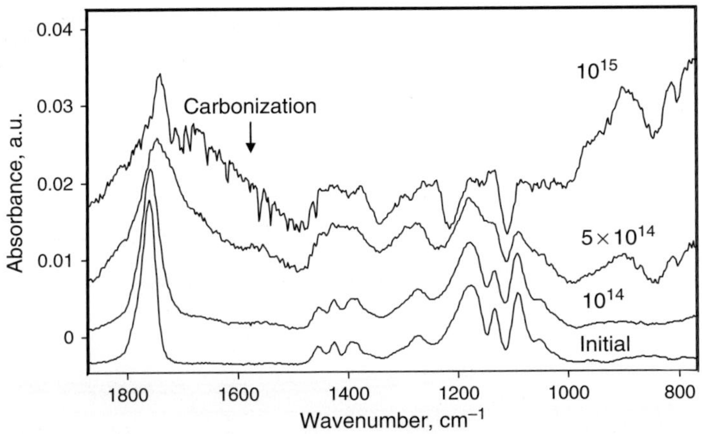

Fig. 10.10. FTIR spectra of PLGA after PIII with 20 keV ion energy (fluence of treatment is noted in ions/cm^2). The spectra have been normalized by the 1380 cm^{-1} line of PLGA. Experiments were done by I. Kondyurina.

of the unmodified PLGA is in the range of 1.47–1.49 over the 400–1000 nm spectral interval. After ion implantation the refractive index increases to 1.6–1.7. At short wavelengths the refractive index for the modified samples is considerably higher than at low wavelengths because of dispersion due to optical absorbance at short wavelengths. The short wavelength absorbance is caused by the appearance of unsaturated carbon structures in the ion modified PLGA (Fig. 10.11). Spectral changes correspond to carbonization of the surface layer under ion beam. The thickness of the PLGA film decreases due to etching of PLGA film under ion beam. The rate of etching depends on energy and fluence of ion beam implantation: at high fluence the effect of carbonization decreases the etching rate. The PLGA film after modification becomes crosslinked and it cannot be dissolved in previously active solvent like acetone. FTIR spectra of insoluble PLGA film shows that the structure of PLGA remains after ion beam.

The crosslinking of PLGA decreases the mobility of the macromolecules and the surface layer becomes much more

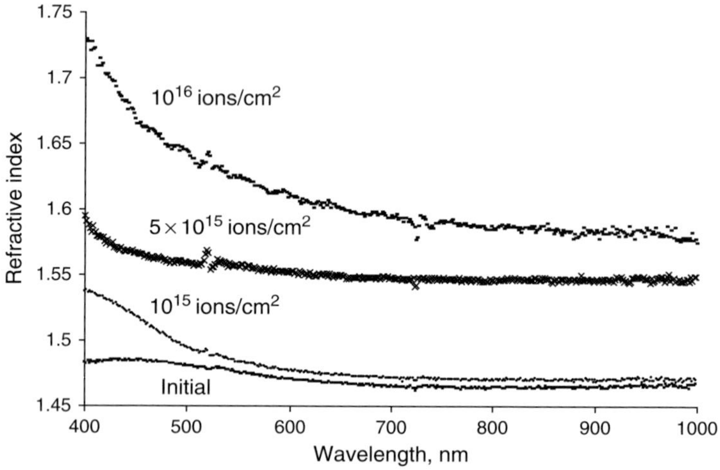

Fig. 10.11. Refractive index of PLGA film after PIII with different fluences of ion implantation. Experiments were done by I. Kondyurina.

stable. The modified film does not rupture in water solution, but the degradation propagates in the film (Fig. 10.12). The final stage of degradation depends on fluence of ion beam implantation. At high fluence (10^{15} ions/cm^2 and higher) the PLGA film is crosslinked and highly carbonized up to formation of a stable non-degradable structure. It is no longer a biodegradable polymer, but a carbonized coating on the substrate. At low fluence (at 10^{14} ions/cm^2 and lower) the crosslinking and carbonization are not so high and the film remains biodegradable. In our experiment we observed 90% biodegradable fraction of the film.

In all cases after ion beam implantation with low and high fluences, the dewetting effect is not observed for thin film, when the thickness of the film is close to ion penetration depth. For thick film, the thickness is much higher than for ion penetration depth, the dewetting process has complex characteristics: the top layer of the film is crosslinked and forms a continuous film (Fig. 10.13). The bottom unmodified layer is not crosslinked and such a layer dewets under the crosslinked layer. We observed

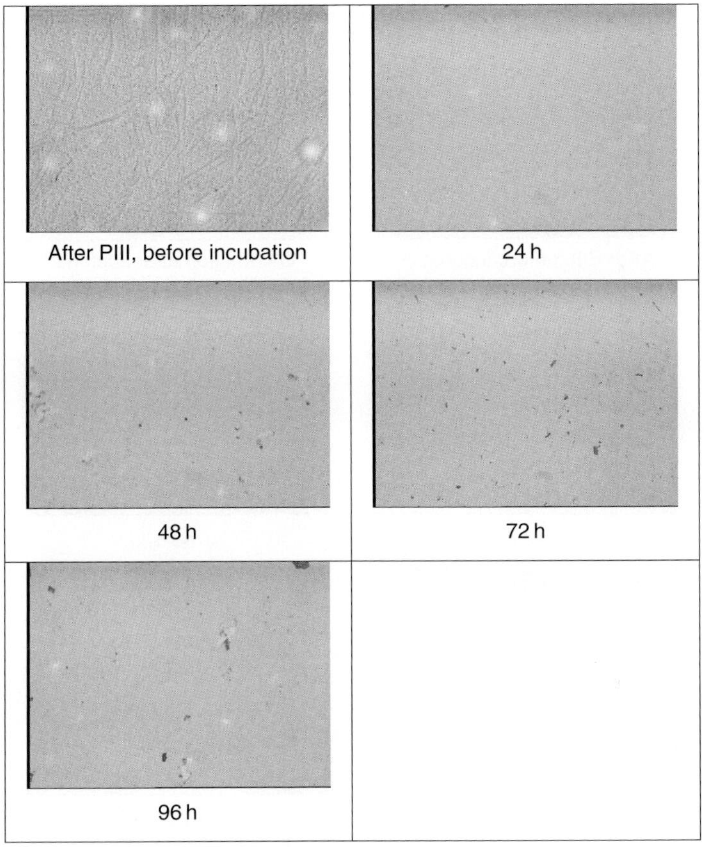

Fig. 10.12. Microphotos of 100 nm PLGA coating after PIII treatment and with time of degradation in water. Size of photos is $1200 \times 950\,\mu m$. Experiments were done by I. Kondyurina.

such underlayer dewetting for PLGA film of 1000 nm thickness, when ion penetration depth was 40 nm. Modification of the thick film requires higher energy of ions and a deeper modified layer of the film for complete cessation of the dewetting process.

Here we consider only convenient ion beam modification of polymers without deposition techniques. PIII and ion beam could be used for assistance of deposition of inorganic

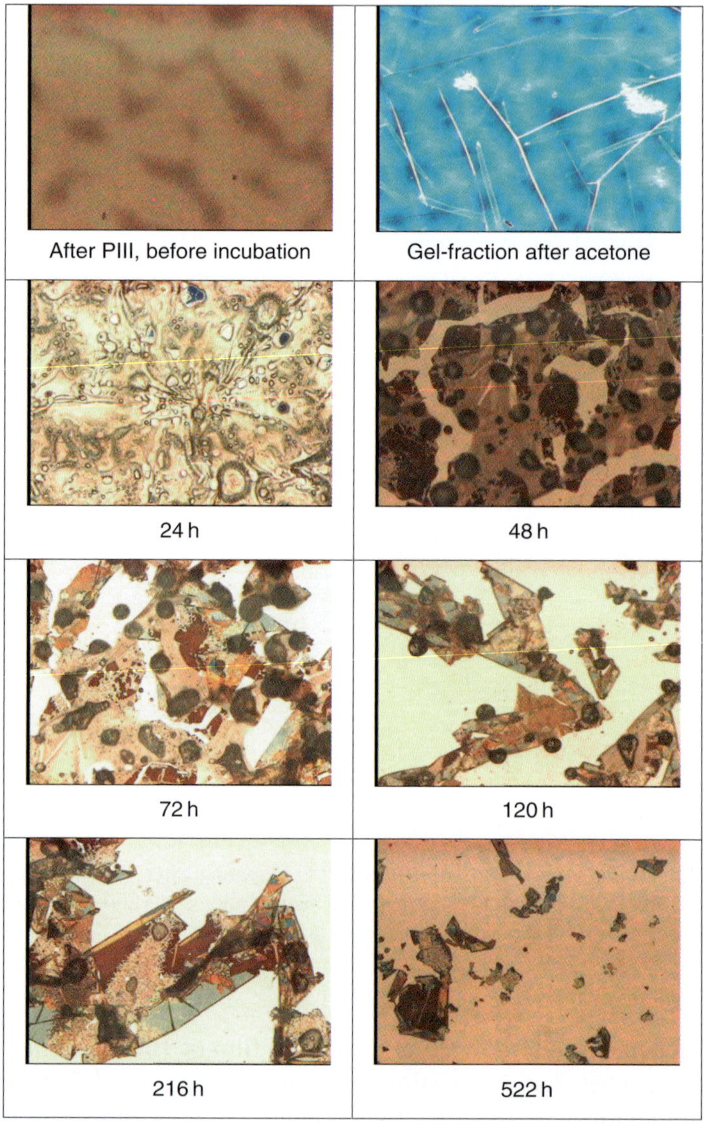

Fig. 10.13. Microphotos of 1 μm PLGA coating on silicon substrate after PIII treatment and with time of degradation in water. Size of photos is 1200 × 950 μm. Experiments were done by I. Kondyurina.

layers on top of polymers to create a preserve coating against aggressive media. Some literature on ion beam modified polymers at degradation in the space environment can be found in [16, 17].

References

1. B. Ranby, J.F. Rabek, *Photodegradation, Photooxidation and Photostabilization of Polymers*, Wiley, London, 1975.
2. D. Briassoulis, A. Aristopoulou, M. Bonora, I. Verlodt, Degradation characterisation of agricultural low-density polyethylene films, *Biosys. Eng.* 88(2) (2004), 131–143.
3. H. Al-Madfa, Z. Mohamed, M.E. Kassem, Weather ageing characterization of the mechanical properties of the low density polyethylene, *Polym. Degr. Stab.* 62 (1998), 105–109.
4. N. Hassini, K. Guenachi, A. Hamou, J.M. Saiter, S. Marais, E. Beucherd, Polyethylene greenhouse cover aged under simulated sub-Saharan climatic conditions, *Polym. Degr. Stab.* 75 (2002), 247–254.
5. L. Kuepper, J.V. Gulmine, P.R. Janissek, H.M. Heise, Attenuated total reflection infrared spectroscopy for micro-domain analysis of polyethylene samples after accelerated ageing within weathering chambers, *Vibr. Spectr.* 34 (2004), 63–72.
6. H. Hinsken, S. Moss, J.-R. Pauquet, H. Zweifel, Degradation of polyolefins during melt processing, *Polym. Degr. Stab.* 34 (1991), 279–293.
7. A.S. Shklyaev, V.A. Balkov, *Climate of Perm Region*, Perm Book Publisher, Perm, 1963.
8. T.R. Crompton, *The Analysis of Plastics*, Pergamon, 1984, 445 pp.
9. H.F. Mark (Ed.), *Encyclopedia of Polymer Science and Technology*, 3rd edn, Wiley, 2004.
10. J.H. Kim, A. Taluja, K. Knutson, Y.H. Bae, Stability of bovine serum albumin complexed with PEG-poly (L-histidine) diblock copolymer in PLGA microspheres, *J. Control. Rel.*, 109 (2005), 86–100.
11. A. Beletsi, Z. Panagi, K. Advoustakis, Biodistribution properties of nanoparticles based on mixture of PLGA with PLGA-PEG diblock copolymers, *Int. J. Pharm.* 298 (2005), 233–241.

12. V. Lemaire, J. Belair, P. Hildgen, Structural modeling of drug release from biodegradable porous matrices based on a combined diffusion/erosion process, *Int. J. Pharm.* 258 (2003), 95–107.
13. R. Yang, T. Chen, H. Chen, W. Wang, Microfabrication of biodegradable (PLGA) honeycomb-structures and potential applications in implantable drug delivery, *Sensors and Actuators* B106 (2005), 506–511.
14. Y. Mo, L.-Y. Lim, Paclitaxel-loaded PLGA nanoparticles: potentiation of anticancer activity by surface conjugation with wheat germ agglutinin, *J. Cont. Rel.* 108 (2005), 244–262.
15. B.B. Crow, A.F. Borneman, D.L. Hawkins, G.M. Smith, K.D. Nelson, Evaluation of in vitro drug release, pH change, and molecular weight degradation of poly(L-lactic acid) and poly(D,L-lactide-co-glycolide) fibers, *Tissue Engineering* 11 N7/8 (2005), 1077–1084.
16. J. Kleinman, Z. Iskanderova, R.C. Tennyson, Ion implantation protects surfaces, *Adv. Mat. Proc.* 4 (1998), 26–30.
17. M. Ueda, I.H. Tan, R.S. Dallaqua, J.O. Rossi, J.J. Barroso, M.H. Tabacniks, Aluminum plasma immersion ion implantation in polymers. *Nucl. Instr. Meth. Phys. Res.* B 206 (2003), 760–766.

11 Polymerization of liquid polymer matrix in free space environment

Space flight is wonderful accomplishment for our civilization. Humankind has a 50-year history of space flights. However, up to now these have been limited to near Earth space. The International Space Station (ISS) is a bigger version of the Spacelab, Salut and MIR stations and we await the next new step in space exploration.

Recently businesses and governments have shown a renewed interest in space activity. This is caused by new technologies which present new possibilities for industry in specific space environments. The private space carrier "Spaceship One" and the first private space hotel from Bigelow Aerospace have been launched. On the other hand, the space environment is a source of new possibilities for businesses: microgravity for biotechnology, extra pure vacuum for semiconductors, and mines on other planets. The American Space Agency NASA, the European Space Agency (ESA), the Russian Space Agency (RSA), the Japan Space Agency (NASDA) and the Chinese Space Agency have expressed interest in large space constructions which can pave a new way for the space industry.

The problems of a lengthy stay in space, deep space flights, orbital plant and Moon or Mars colonies are compounded by the size and mass of space construction, and its delivery from Earth surface to Earth orbit. However, the size and mass of space construction sent to Earth orbit are limited by the size of the launch vehicle.

The problem of making large-size constructions can be solved by using the technology of the polymerization of fiber-filled composites and a reactionable matrix applied in free space or on another space body when space construction takes place over a long period of time. For the creation of the construction frame the fabric impregnated with a long-life matrix (prepreg) is prepared in terrestrial conditions and, after folding, can be kept on board a spaceship after launch. In due time the prepreg is sent into free space and unfolded by, for example, inflating. Then a reaction of matrix polymerization takes place. After polymerization, the durable frame can be fitted out with air under normal pressure. Next, the crew can carry out work in the frame without space suits, to build a station, space plant, space hotel or greenhouse, for example, fitting the frame with apparatus and life support systems.

In this case, frame size and the form of future space construction are not limited by the ability of modern space carriers. The launch of a bulky frame for a large-size space station, solar cell panel or antenna needs one space carrier, reducing any form of risk at lift-off. The creation of the frame does not require dangerous and complicated procedures to join separate launch vehicles in orbit and provides otherwise extremely difficult hermeticity of the joints.

However, the technology of polymerization was not used for creation of space construction. The problem of polymerization in free space is connected with specific conditions of the free space environment for polymer materials. The conditions of free space have a great destructive influence on polymer materials, especially on the liquid polymer matrix. In the free space the composite material is treated with a high vacuum, sharp temperature changes, plasma of free space formed by space rays, sun irradiation and atomic oxygen (on low Earth orbit), micrometeorite fluency and microgravitation [1–8]. Free space conditions can be simulated on Earth in a high vacuum chamber, plasma discharge and ion beam implanter. In this chapter we consider the polymerization processes in simulated free space environment conditions of low earth orbit (LEO) and geostationary earth orbit (GEO) which are of most interest for space constructions.

Vacuum

The altitude of LEO varies from 90 to 1000 km. The usual orbit for most space flights varies between 300 and 400 km in altitude. The pressure of residual atmosphere on an altitude of 300 km is reported as 10^{-3}–10^{-5} Pa. The pressure near the spaceship or space construction depends on time from launch. During flight desorbing gases, venting trapped volumes, releasing dust and ice particles can increase the pressure near a new space construction. By experimental observation the pressure can vary from 10^{-5} to 10^{-3} Pa depending on the time of flight, sun irradiation, local configuration and spaceship materials, and activity of the engine. Even at a distance from Earth of about 36 000–42 000 km corresponding to a GEO mission, the pressure would not equal 10^{-9}–10^{-11} Pa. Due to continuous evaporation the pressure near a satellite or spaceship would be significantly higher.

The temperature in space experiments in Earth orbits usually varies from −150°C to +150°C and even 200°C. The combination of high vacuum and high temperature influences dramatically the evaporation process of polymer materials. The main effect of a such vacuum is observed as evaporation of low molecular mass fractions. The rate of evaporation into a high vacuum is described by Langmure's formula:

$$W(\mathrm{g}/(\mathrm{cm}^2 \cdot \mathrm{sec})) = (P/17.14) \cdot (M/T)^{0.5}$$

where M is the molecular mass of vapor, T is the temperature (K), and P is the equilibrium vapor pressure of fraction from the Klausius–Klapeyron equation [9]. With a lengthy vacuum action the stoichiometry of low weight molecular components in the matrix can be changed [10].

Space plasma

Space plasma is created by the fluency of galactic and Sun protons, electrons, neutrons and heavy particles with wide diapason of energy from some eV to hundreds of MeV; infrared, visual, ultraviolet and vacuum ultraviolet photons and X-ray photons. On LEO the fluence of atomic oxygen (AO) is added as a more significant factor in comparison with other factors of space plasma on GEO [11]. The main destruction factors of space plasma for hard polymer materials by space experiments and laboratory experiments are thermal atomic oxygen, high energy electrons and protons flows, VUV irradiation and X-ray irradiation. The parameters of space plasma in space construction depend on space weather: sun activity, Earth atmospheric events, day/night time, parameters of space orbit, Earth position and space construction conditions. Due to variations of all these conditions, the space plasma parameters can vary in 100 times frame of value, which was measured in different space experiments and flight missions.

Atomic oxygen

Via experiments on LEO with degradation of polymer materials the AO is the main factor which limits the exploitation time of materials. The estimation of average AO flux on LEO (near 300 km altitude) during real experiments with polymer materials gives $2.88 \cdot 10^{13}$ at/cm^2/s average value in [11], $3.88 \cdot 10^{13}$ at/cm^2/s in LDEF mission [12], 10^{14} in MISSE mission [13], $5 \cdot 10^{14}$ at/cm^2/s theoretical value and $4.3 \cdot 10^{14}$ at/cm^2/s on Kapton equivalent for ESEM mission [14], $5 \cdot 10^{13}$ at/cm^2/s in [15] and 10^{12}–10^{15} at/cm^2/s in [16] for ESEM mission too, 10^{13}–10^{15} at/cm^2/s in [17]. In the Habble mission (595 km altitude) the AO flux equaled $6.86 \cdot 10^{11}$ at/cm^2/s. The flux of AO varies due to Sun activity, season, position, longitude/latitude and altitude of spaceship, variations in Earth atmosphere and outgassing processes of spaceship materials. The model approximation developed in [18] shows an AO flux of $2 \cdot 10^{11}$ atoms/cm^2/s at 400 km altitude.

Vacuum ultraviolet irradiation

Vacuum ultraviolet (VUV) irradiation is part of the solar spectra of irradiation. The intensity of VUV light is low, but the effect of VUV light on polymers is significantly higher than visual and UV light. At LEO the level of VUV light was estimated of about $4 \cdot 10^{11}$ photons per cm^2 per second for a 121.6 nm wavelength [19]. The Sun irradiation density corresponds to 0.75 mkW/cm^2 in a VUV diapason of 100–150 nm wavelength in [20] and 11 mkW/cm^2 in a UV diapason of 200–300 nm wavelength in [21].

A number of gas-discharge sources are used for the simulation of solar VUV light in laboratory experiments. The hydrogen lamp generates a line at 121 nm in a VUV diapason. The deuterium lamp generates a continuous spectra in a diapason of 115–400 nm. Krypton and argon lamps are used too. Usually, the irradiation intensity of some units of solar VUV activity from lamps is used for rapid ageing of polymers. The VUV factor is significant for polymers at LEO and GEO missions.

X-rays

The level of X-rays in Earth orbit equals $2.3 \cdot 10^{-9}$ W/cm^2 for a 1–8 Å wavelength and $1.43 \cdot 10^{-10}$ W/cm^2 for 0.5–4 Å wavelength [22]. The majority of flux from X-rays is directed from the Sun with less from other stars.

The simulation of X-ray irradiation was made with a Cu K_{λ} line source with electron excitation energy lower than 60 keV. The intensity of X-ray irradiation was $1.3 \cdot 10^{-3}$ W/cm^2.

High energy particles

The energetic spectrum of electron and ion fluxes at LEO and GEO is sufficiently complex. The energetic spectrum and flux of charged particles depends on the kind of particle, altitude, longitude/latitude, seasons and Sun activity. The energy of charged particles varies in diapason from 0.1 eV to some GeV. In laboratory experiments some theoretical models are used for the analysis of effects in materials generated by charged particle flows. The density of electrons at an LEO altitude of 400 km equals 10^5 e/cm^3 (night side) and 10^6 e/cm^3 (day side) in [11] with energy of 0.1 eV in [19]. The electron flux in a GEO mission equals 10^9 e/cm^2/s for electrons with energy of 0–12 keV in [23]. The electron density in a GEO mission of 1.12 e/cm^3 at average energy of $1.2 \cdot 10^4$ eV and ion density of 0.236 ion/cm^3 with average energy of $2.95 \cdot 10^4$ eV is used for the simple analysis of plasma in GEO missions [24]. The most frequent ions are hydrogen ions (90%). The other 10% of ions consist of all elements.

Laboratory simulation of high energy particles is made on the basis of accelerators of electrons and ions. The energy of electrons in Earth laboratory experiments [21, 25, 26] varies from 20 keV to 2 MeV. The energy of ions in experiments varies from 40 keV in [25] to 2 MeV in [21]. Principally, the electron and ion beams are used for the simulation of the GEO space environment on polymers.

Temperature variations

The total solar irradiation from the Earth to the Sun equals 1362–1367 W/m^2 in [11]. The solar irradiation level depends on the season (position of Earth on solar orbit) and it can vary from 1316 W/m^2 at minimal solar energy flux (summer solstice) to 1428 W/m^2 at maximal solar energy flux (winter solstice) in [18]. The level of de-irradiation of sunlight by the Earth's surface and its atmosphere equals 240 W/m^2. Temperature changes in the spaceship frame depend on its orientation to sunlight, absorption and emission indexes of the frame surface and internal heat sources [27]. The real temperature of the frame surface in LEO can vary in wide diapason: $-56 \div +77$°C in [28], $-90 \div +120$°C in [29], -100°C $\div +200$°C in [30] and [22], and $-150 \div +150$°C in [31]. For far space missions such as the NGST mission (halo orbit, $1.5 \cdot 10^6$ km from Earth) the estimated temperature equals $-223 \div +122$°C in [32].

In LEO the surface of the spaceship is under thermal cycling for a period of about 90 min due to sunlight and Earth shadow. In GEO the surface temperature of the spaceship depends only on orientation and rotation of the spaceship in relation to sunlight. Thermal cycling has a destructive effect on polymer materials. The analysis of temperature changes at rotation of the frame to the Sun showed that the appearance of a sharp temperature gradient in the space station frame during curing time is possible. The appearance of a temperature gradient in curing plastic leads to the formation of internal stresses in the cured frame after decreasing or changing the sign of the temperature gradient [33].

Microgravity

On LEO and GEO missions a spaceship exists in microgravity conditions. The level of gravity is two orders lower in comparison with gravity on the Earth's surface. The gravity level on the spaceship depends on acceleration conditions of the spaceship, light pressure, gravitation gradient, movements of the crew, docking, etc. The polymerization of acrylamide gel was made during space flight on the Shuttle [34] and on the "Mir" space station [35, 36]. Polymerization was held inside the spaceship, the influence of microgravity on polymerization process was investigated [37]. Other factors of space environment were excluded. Also, the influence of microgravitation on the polymerization process was studied in experiments of downfall at the Drop Tower of the

Bremen Centre of Microgravity [38] and during the flight of sounding rockets by investigators from the University of Southern Mississippi [39].

The influence of microgravity on polymerization is based on the exclusion of convection and sedimentation processes in the curing polymer [1, 40, 41]. Under microgravity conditions a frontal polymerization process and creation of highly homogeneous polyacrilamide gel were observed [40–43]. In high filled polymers, such as epoxy composite materials, the microgravity should not influence the polymerization process. However, for tall constructions of prepreg there is the limitation of high dimension due to flow of liquid resin in the Earth's gravity field. By estimation, the limit in height for constructions with liquid resin on Earth is about 3 m. In microgravity in LEO and GEO the limit for large-size construction dimensions due to flow of liquid resin should be absent.

Meteorite fluency

Meteorite fluency leads to the mechanical destruction of material. In the case of liquid polymer the influence of meteorite fluency is weaker than for a hard polymer, because a collision of meteorite particles with the surface of liquid polymer does not lead to crack formation. Erosion of polymer at meteorite fluency corresponds to 0.1 nm/year in far space and 20 nm/year on low Earth orbit [9].

The conditions in the free space environment are extreme for solid polymer materials. Destruction of hard polymers in free space was studied on many occasions, including real experiments outside the spaceship in LEO, GEO and in laboratory modeling experiments of space action on Earth (see, for example, [13, 17, 19, 20, 26, 29, 31, 32, 44–74]). These studies showed that destruction of epoxy composites, HDPE, LDPE, Kapton, Teflon and FEP Teflon, Tedlar, Mylar, fluorinated polyimides, polyurethanes, polysulfone and others, during flight in low Earth orbit is connected with plasma action of atomic oxygen with low kinetic energy. The basic results of destruction are loss of mass, dehydration, formation of an amorphous carbon layer, formation of cracks and craters on surfaces and a decrease in the durability of polymer.

In laboratory experiments the influence of separate space factors on polymers was studied in more detail. The conditions of the simulated free space environment used for polymer degradation in the laboratory can be overviewed in Table 11.1. The main influence at LEO mission is connected with AO and VUV actions. The main influence at

Table 11.1. Free space environment simulated conditions for experiments on polymerisation

Factor of space environment	LEO	GEO
Temperature	$-150 \ldots +150$°C	$-150 \ldots +150$°C
Vacuum	$10^{-1} \ldots 10^{-3}$ Pa (*)	$10^{-1} \ldots 10^{-3}$ Pa (**)
Atomic oxygen	$10^{12} \ldots 10^{15}$ atom/cm^2/s	–
UV light	0.75 μW/cm^2 at 100–150 nm wavelength	0.75 μW/cm^2 at 100–150 nm wavelength
	11 μW/cm^2 at 200–300 nm wavelength	11 μW/cm^2 at 200–300 nm wavelength
X-rays	10^{-3} W/cm^2, Cu K_α, 60 keV	10^{-3} W/cm^2, Cu K_α, 60 keV
Electron beam	10^9–10^{12} electron/cm^2/s at 40 keV energy	10^9–10^{12} electron/cm^2/s at 40 keV energy
Proton beam	10^5–10^9 proton/cm^2/s at 40 keV energy	10^5–10^9 proton/cm^2/s at 40 keV energy

(*) The variation of the values corresponds to the variation of solar activity with time.
(**) The pressure of $10^{-9} \ldots 10^{-11}$ Pa was estimated for GEO missions without any evaporation from spacecraft material. Real pressure is higher ($10^{-3} \ldots 10^{-5}$ Pa). A pressure lower than 10^{-3} Pa for high evaporated substances like polymers is not achieved by known vacuum pumps. However, a pressure of $10^{-1} \ldots 10^{-3}$ Pa is lower than an evaporation pressure of known monomers and oligomers for polymer matrix. Therefore a pressure of $10^{-1} \ldots 10^{-3}$ Pa in laboratory experiments is equivalent to real pressure at LEO and GEO missions.

GEO mission is connected with VUV, electrons and ion erosion actions. The AO, electrons, ion fluxes and VUV intensity depend on Sun activity and spaceship mission conditions. NASA Guidelines with "The Space Environments and Technology Archive System" at LaRC, Boeing Database and Canadian Database can be used for simulation of space orbit conditions during virtual space missions on hard polymer materials.

The effects of degradation can be observed using the following methods: optic and electron microscopes and AFM reveal an increase in the roughness of the polymer surface; XPS spectroscopy reveals oxidation of the surface layer; and mechanical and thermo-mechanical methods show a decrease in durability and an increase in modulus and T_g. Comparison of polymer degradation results in real free space experiments and in simulated free space experiments shows similar changes of structure and characteristics of polymers.

An investigation of the polymerization processes in liquids or oligomers under real space conditions during space flight was not found in literature. The process of epoxy resin cure includes a chemical reaction of polycondensation, when an epoxy ring of the macromolecule opens under the action of a catalytic agent or active group of hardening agents with the formation of chemical bonds between macromolecules. During curing the molecular mass of components increases up to the formation of a "physical" or "chemical" polymer network. The viscosity increases slowly at the first period of reaction and extremely quickly at the second period of reaction. The kinetic process of reaction is described by Arrhenius' law with acceleration and deceleration stages of reaction. The kinetic equations contain concentrations of active components and the evaporation of active components influences the kinetic behavior of reaction.

Usually, long lifetime epoxy compositions are considered for space application in deployment structures. In this case the curing reaction must be initiated at the right time.

The curing of epoxy resin can be initiated by the heating of a reactionable mixture. This is the usual way for initiation of chemical reactions. The lifetime of such systems depends on temperature. For a long lifetime the mixture must be saved at low temperature and/or heated to high temperature for polymerization. In this case the thermal regime of construction is important to manage the curing reaction in space.

The kinetic reaction can be initiated by irradiation of UV or visual light. The special UV photoinitiators are inserted into epoxy resin. The UV photoinitiator initiates the curing reaction under UV light. The lifetime of such a system is very long and depends on the level of light at storage. The polymerization process in epoxy composition during UV curing corresponds to the frontal process of moving the polymer/oligomer boundary into bulk layers of material.

Let us consider curing a polymer matrix, which was investigated for the polymerization process in a free space environment.

Polymerization reaction in modeling composites is carried out with amine hardening agents of triethanolamine (TEA) and triethylenetetramine (TETA). With TETA the condensing reaction leads to the formation of hydroxyl groups and secondary amine groups. This reaction is sensitive to stoichiometric ratio of components and this reaction has a wide application for the synthesis of epoxy matrix in carbon and glass reinforced composites. The reaction with TETA hardener agent proceeds at room temperature during 6–8 hours which is useful for modeling studies.

The reaction of epoxy resin with TEA hardening agent proceeds at a high temperature and the composition can be saved in the liquid state after mixing over a number of days. This reaction is useful for modeling studies when a time of non-curing reactionable system should be used for pretreatment of the mixture before polymerization. The reaction of epoxy resin with TEA was used for modeling the polymerization process under specific conditions. The familiar way for writing this reaction is:

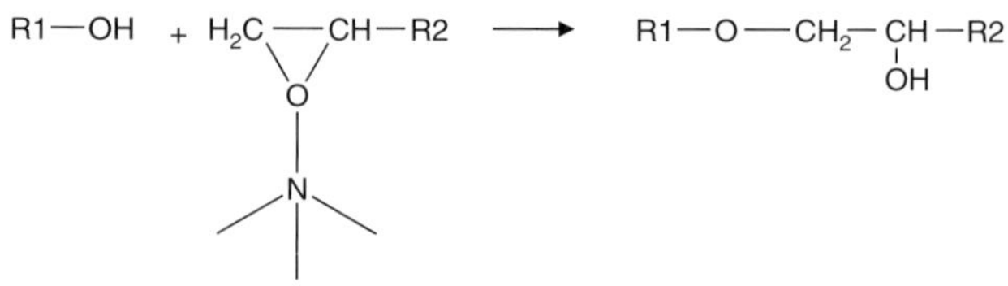

The kinetics of the polymerization reaction is well observed by FTIR and Raman spectroscopy. For example, FTIR spectra of epoxy resin with TETA hardening agent during curing is shown at Fig. 11.1. The intensity decrease of the 915 cm^{-1} line shows the decrease of the epoxy group concentration during reaction. The intensities of lines at 862, 971, 1078, 1347, 1429, 3000 and 3056 cm^{-1} related to epoxy and ether groups show the reaction kinetics also. Usually the 915 cm^{-1} line intensity is used for quantitative analysis of reaction.

Raman spectra of the reaction mixture are useful for reaction analysis of thick composite samples in experiments with vacuum, plasma and ion beam treatment when thick composite samples cannot be analyzed by FTIR spectra. In this case for Raman spectra the sample is irradiated locally in small point with diameter of laser beam. Raman spectra of epoxy resin with TETA hardening agent during curing are shown in Fig. 11.2. Analysis of reactive groups can be done by intensity of lines at 677, 770, 799, 987, 1053, 1134, 1160, 1260, 1430 and 1482 cm^{-1}. Time dependence of these lines' intensity and FTIR spectra are used for calculating the stage of the curing reaction in the epoxy matrix.

The analysis of the polymerization kinetics of epoxy resin and TEA hardening agent at different temperatures is presented in Fig. 11.3. The dependence of reaction stage on

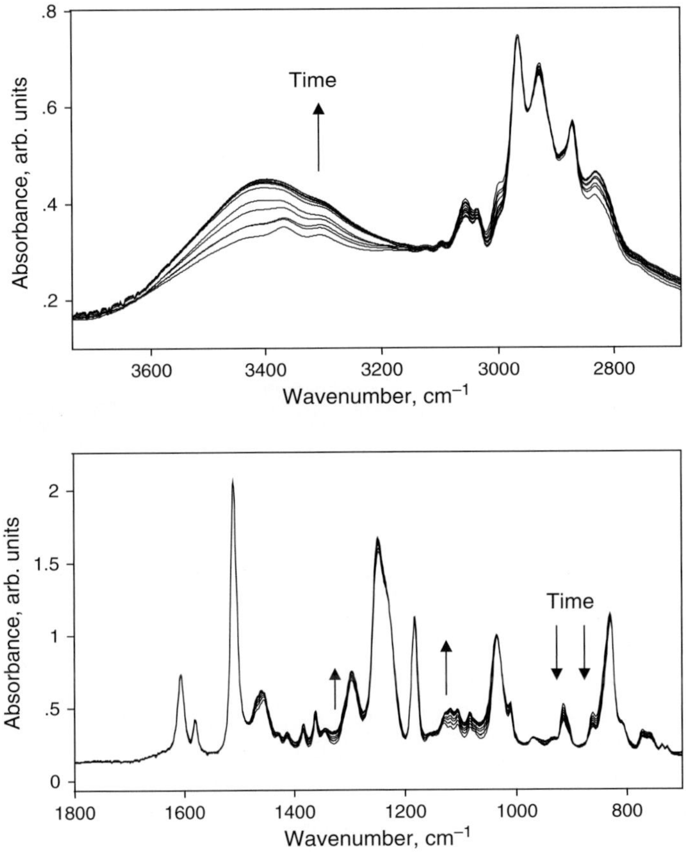

Fig. 11.1. FTIR spectra of epoxy resin with TETA hardening agent during curing.

time corresponds to the type of polymerization reaction of second order with autocatalyst and autodeceleration:

$$\frac{\partial \beta}{\partial t} = k_1 \cdot (1 - a\beta) \cdot \left(\beta + \frac{k_2}{k_1} \right),$$

where β is the stage of polymerization, which is calculated from intensity of active group lines in FTIR or Raman spectra.

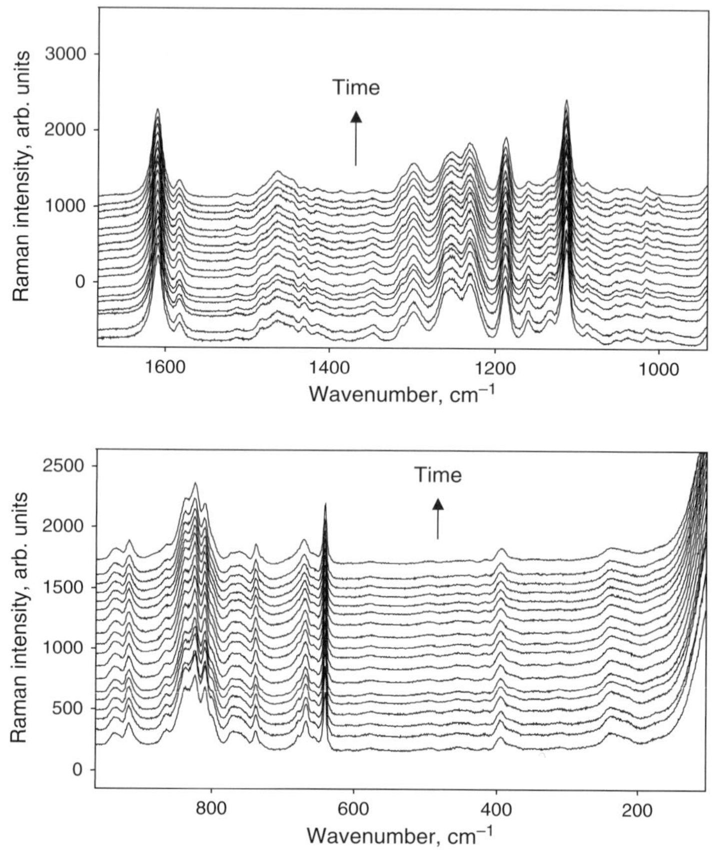

Fig. 11.2. Raman spectra of epoxy resin with TETA hardening agent during curing.

The curing kinetics is observed using the viscosity of reactionable mixture of epoxy resin and hardening agent. With time the viscosity of reactionable mixture increases, which corresponds to the polymerization reaction. The matrix viscosity increase with time of the reaction is responsible for an increase of modules of filled composite. The curves of the epoxy group concentration (by FTIR and Raman), viscosity of mixture and modulus of composite with time show the kinetics of polymerization (Fig. 11.4).

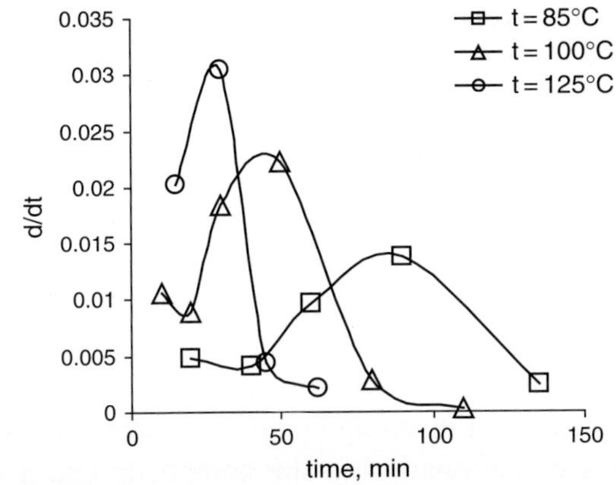

Fig. 11.3. Curing rate derivative on time at different temperature for epoxy resin with TEA hardening agent.

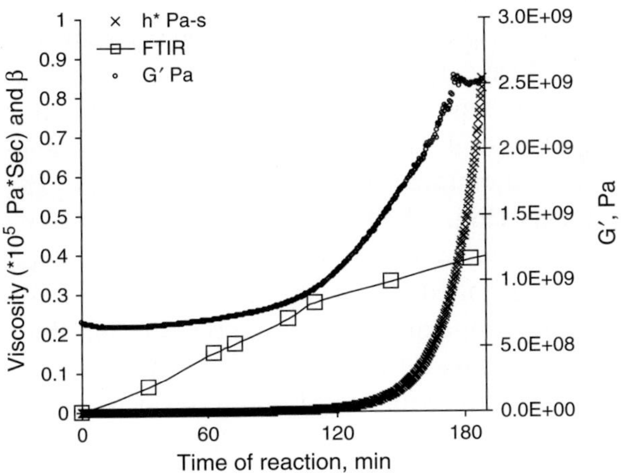

Fig. 11.4. Viscosity of resin, module of glass fiber composite and stage of polymerization during curing for epoxy resin with TETA hardening agent.

The vacuum of free space in LEO and GEO orbits ($10^{-3} \ldots 10^{-5}$ Pa) is too high in comparison with vapor pressure of all components of the polymer matrix. In high vacuum the stoichiometric ratio of active components in the liquid matrix changes due to different evaporation rates of low molecular weight components. The evaporation process of separate components depends on the vapor pressure of components, diffusion coefficients of components in the polymer matrix, thickness of the matrix layer, temperature and external pressure. In the case of fractional constitution of the epoxy resin and hardening agent, the low weight fractions can quickly evaporate and the chemical activity of the mixture can decrease with time. The deviation of stoichiometric ratio and the decreasing of chemical activity of the matrix can cause the polymerization reaction to stop. The evaporation of liquid components can decrease the polymer matrix contamination in the composite and a ratio of matrix/fiber contamination can shift the optimal ratio for expected properties of the composite.

The evaporation of individual polymer matrix components was analyzed at various pressures and temperatures. For example, the mass loss of epoxy resin based on Bisphenol A, average MM = 374, is presented in Fig. 11.5. An increase in temperature accelerates the evaporation process. A decrease in pressure increases the evaporation rate. High evaporation rate is observed for TETA and TEA hardening agents (Fig. 11.6). TETA evaporates at room temperature in vacuum. At high temperature and low pressure TETA can be boiled. The evaporation of TEA is observed only at high temperature and low pressure, but the rate of evaporation is higher than for epoxy resin.

The evaporation kinetics of epoxy resin is complicated. The resin consists of a number of fractions with different pressures of vapor and different rates of evaporation. The kinetics rate of epoxy resin evaporation decreases with time. At first, light fractions evaporate with high rate of evaporation, and then heavy fractions evaporate with low rate of evaporation. A simulation of the evaporation process from a mixture of components with different evaporation rates is presented in Fig. 11.7.

If evaporation of a hardening agent is strong enough, the stoichiometric ratio becomes broken during polymerization under vacuum. Experimentally such a situation is

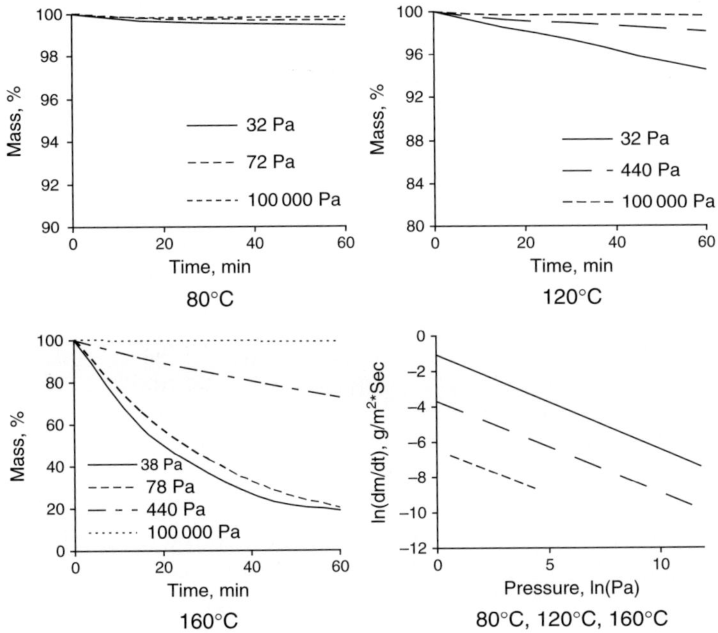

Fig. 11.5. Evaporation kinetics of epoxy resin ($M_w = 374$) without hardening agent at different temperatures and pressure.

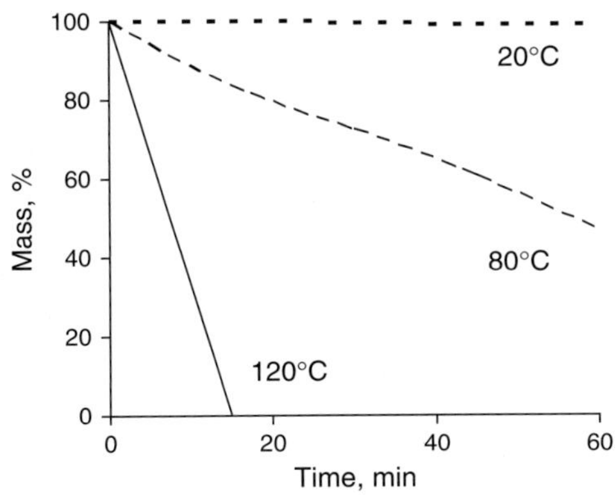

Fig. 11.6. Evaporation of TEA hardening agent at 30 Pa pressure and different temperature.

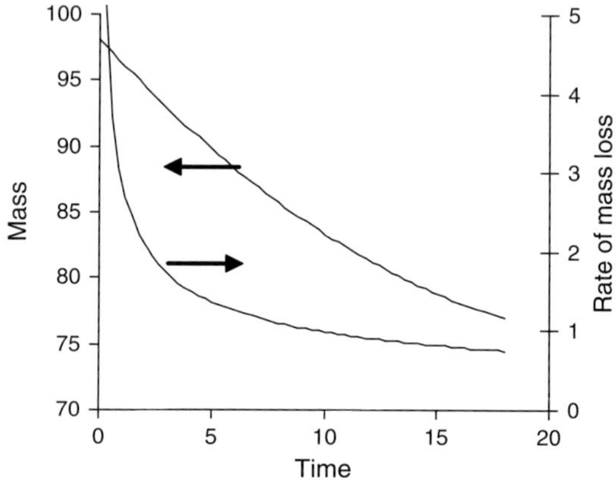

Fig. 11.7. Simulated evaporation process of multicomponent liquid mixture with different individual rates of evaporation.

observed for an epoxy matrix based on epoxy resin with an isomethyl-tetrahydro-phthalate anhydride hardening agent (Fig. 11.8). The polymerization reaction with anhydride hardener proceeds at a high temperature of 130–150°C. However, the evaporation of anhydride proceeds under low temperature and sufficiently high pressure. At low pressure

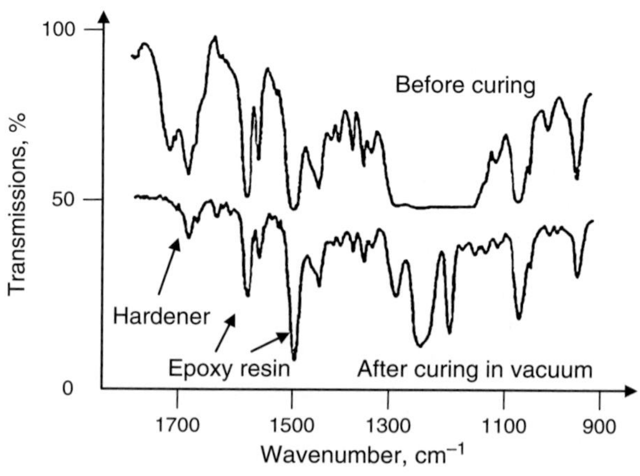

Fig. 11.8. FTIR spectra of composite based on epoxy resin with anhydride hardener certified for aerospace applications before and after curing in vacuum.

and high temperature closed to real space conditions, the evaporation of anhydride proceeds even with boiling. Therefore, the hardener evaporates more quickly than the epoxy resin. Before the reaction the spectrum shows lines of epoxy resin and hardener. After exposure under vacuum and temperature of polymerization, the spectrum no longer contains hardener lines. The hardener has been completely evaporated during reaction. The epoxy groups have not been spent and the reaction is unfinished. The matrix remains soft.

Intensive evaporation can cause a boiling effect of the liquid, as it is observed in a simple experiment with water under low pressure. The mixture is boiling, the bubbles are formed. After polymerization a hard foam structure of matrix is observed. This breaks the structure of the composite material cured under vacuum and disturbs mechanical properties of the composite. To prevent breaking the composite, any boiling and foaming effects in the liquid matrix during polymerization reaction in a vacuum must be completely excluded.

The measurements of evaporation rates showed that the evaporation process is significant for individual components of polymer matrix even at higher pressure than in free space. The evaporation rates of epoxy resin are significantly lower than for TETA and TEA hardening agents. An exclusion of the evaporation of liquid polymer matrix in free space environment is impossible. However, the negative effect of evaporation can be decreased or excluded.

At LEO missions the atomic oxygen (AO) fluence is the most significant factor of the free space environment on construction materials. During exposition of polymer in AO flow the processes of etching and chemical reactions create the etched oxidized surface layer of polymer. The AO effect on polymers is simulated in plasma discharge. For a curing system a period of liquid composition should be critical due to small molecules of epoxy resin and hardening agent. In this case the bombarding by oxygen atom breaks the chemical bonds of the polymer backbone and short separated parts of destroyed molecules evaporate quickly. With time, during the curing reaction macromolecules become heavier and the breaking of the backbone does not lead to catastrophic degradation and evaporation of the polymer. Therefore, the first period of polymerization reaction is very important for erosion activity of high energy particles. On the other hand, plasma action influences the curing reaction by the way of formation of free radicals and movement of treated layers into bulk layers of liquid composition. In this case, the free radical macromolecules can take part in polymerization reactions with the formation of additional crosslinks in the complete bulk layer of the composite.

The VUV and electron flow penetrating the polymer can generate free radicals too. Such radicals can take part in the curing reaction and increase the curing rate. The kinetic rate of the curing reaction has non-linear dependence on the concentration of active groups in the reactionable mixture. Therefore, the actions of AO, VUV, and electron beam together with the mixing processes of the liquid matrix on the kinetics of the curing reaction cannot be predicted by experiments with separate actions. Taking into account the non-homogeneity distribution of intensity of VUV light and electron density in polymer, the behavior of the polymer matrix during curing depends on the thickness of the matrix and the presence of filler. A combination of different free space environment factors gives an extremely complicated picture of the curing processes in the LEO environment. In this case the real space experiments have a great significance for understanding of space plasma environmental effects.

At GEO missions the high energy electrons and ions flows with VUV irradiation are the main factors of free space environment for polymer degradation. For high energy particles the penetration depth into the polymer is much higher than in AO action. The structure changes of the polymer matrix in bulk layers are available in this case. The flux of high energy particles is lower than AO. During high energy electron and ion irradiation the free radicals can influence the kinetics of polymerization as for the LEO environment. The polymerization process of epoxy composition in the free space environment contains: the chemical reaction kinetics of epoxy groups with hardening agent and epoxy groups with free radicals; reactions between free radicals and free radicals with initial macromolecules; convection processes depending on viscosity; carbonization processes caused by high energy particles with processes of etching; and component evaporation and diffusion from surface layers.

The action of space plasma is very significant at the first stage of polymer presence in space, when the composition is liquid. After polymerization the composition becomes hard and the degradation process under space plasma goes by known way, as it was studied in real space experiments. Therefore, the curing of the polymer matrix initially from liquid to solid states under space plasma environment is key for the development of the polymerization technology of space constructions.

The influence of plasma treatment on uncured epoxy matrix is observed in experiments with liquid epoxy resin. The liquid epoxy resin without hardening agents was put on KBr pellets and treated with oxygen plasma at low pressure. The FTIR spectra of the resin show the changes after exposure in plasma discharge (Fig. 11.9). The optical density of

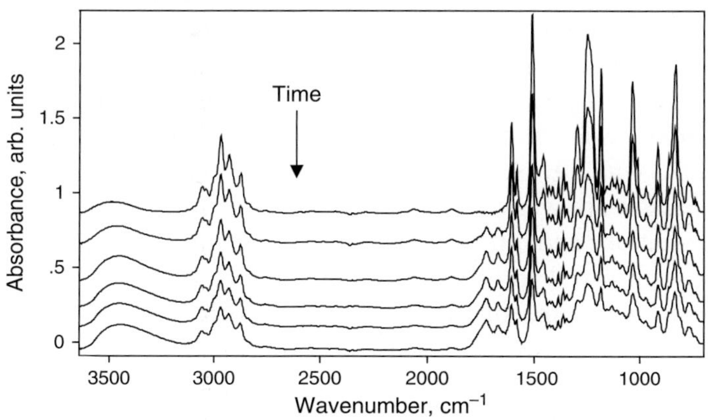

Fig. 11.9. FTIR transmission spectra of liquid epoxy resin with time in plasma discharge.

the line at 1200 cm^{-1} of the ether group, at 1610 cm^{-1} and 1510 cm^{-1} of the aromatic ring of resin, 1450, 1370 cm^{-1} and 2900 cm^{-1} of the methylene groups and 3600 cm^{-1} of the hydroxyl groups decreases with plasma exposure time. These spectral changes correspond to the etching process of epoxy resin in plasma. With the increase of plasma treatment time the thickness of the epoxy resin layer decreases and the optical density of all spectral lines of the epoxy resin decreases too.

If spectra are normalized on the thickness of the resin layer, it is observed that the 1730 cm^{-1} line of the carbonyl group appears and the optical density of this line increases with an increasing in time of the plasma treatment. The increase in this line intensity corresponds to the oxidation processes in epoxy resin under plasma action. This line has a complex profile that corresponds to different carbonyl-containing groups in resin macromolecules after plasma. These are carboxylic acid, ester and aldehyde groups. These groups appear under plasma action during destruction processes of macromolecules of resin. In the hydroxyl stretch vibration region the optical density at 3450 cm^{-1} increases after plasma treatment. The line is wide and interpreted as a stretching vibration line of the hydroxyl groups, which appear as a result of resin destruction in plasma too. The increase in line width in all spectra regions is observed for a long time during plasma treatment.

Similar spectral changes are observed in spectra of cured epoxy resin treated with plasma (Fig. 11.10). The epoxy resin with TETA hardening agent was used. However, the changes are weaker than in liquid resin at the same plasma conditions and time of treatment. The decrease in line intensity observed in all spectra corresponds to etching of the resin with time of plasma treatment. The rate of spectral changes in cured resin is less than for liquid resin. The weak line of the carbonyl group stretching vibrations at 1730 cm^{-1} is observed in the spectra of plasma treated hard resin. The weak increase in absorbance in the region of the hydroxyl group stretch vibrations is also observed. These spectral

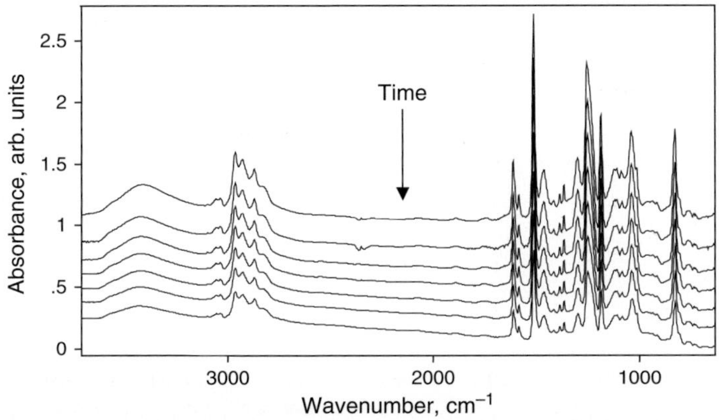

Fig. 11.10. FTIR transmission spectra of solid epoxy resin with time in plasma discharge.

changes are observed at the beginning of plasma treatment and after a lengthy time of plasma treatment the intensity of carbonyl and hydroxyl group lines does not increase. Such spectral changes in hard resin correspond to well-known data on modification of polymers in low pressure plasma. The weak spectral changes are explained by surface character of plasma modification: the thin surface layer of the resin is modified and etched during plasma treatment. The bulk layer of resin is not changed during plasma treatment.

Similar changes in the spectra of liquid epoxy resin with hardening agent are observed during plasma treatment (Fig. 11.11). In these experiments, the reactionable mixture of epoxy resin and hardening agent was placed on KBr pellets immediately after preparation. The decrease in line intensity in all regions of spectra shows the etching processes of composition as in the spectra of liquid and hard epoxy resin. In normalized spectra the chemical structure changes are observed. By the stretching of vibration lines of the carboxyl and hydroxyl groups at 1730 and 3450 cm^{-1} the oxidation process is observed during plasma treatment. The reaction of polymerization is observed by the intensity of the epoxy ring vibration lines at 915 and 860 cm^{-1}, by the

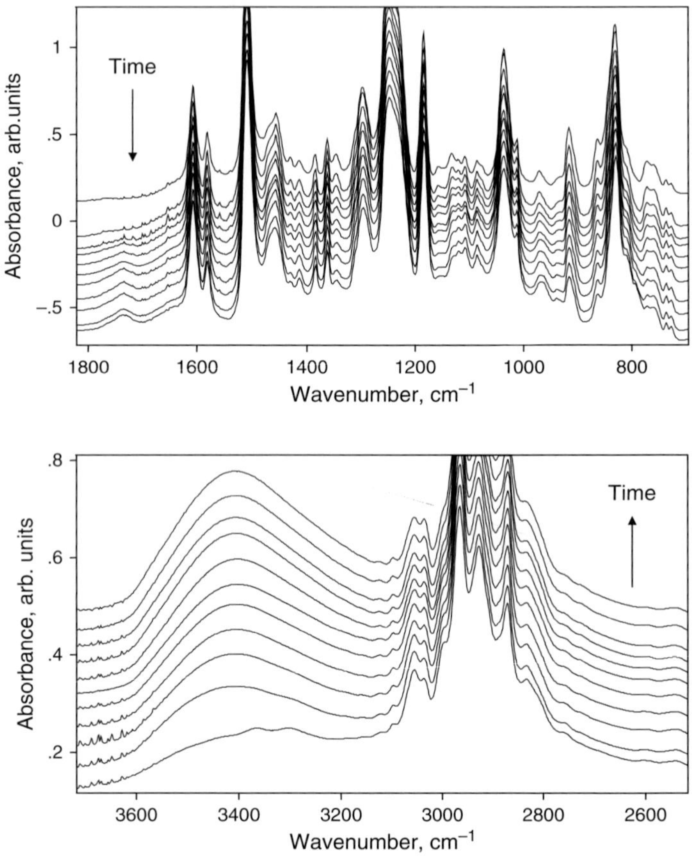

Fig. 11.11. Normalized FTIR transmission spectra of epoxy resin with TETA hardening agent during curing in plasma discharge.

intensity of stretching vibrations lines of the methyl group near the epoxy ring at 3000 and 3057 cm^{-1}, by the intensity of ether group vibration lines in the C—O vibration region at 1260, 1120 and 1107 cm^{-1} and in the O—H stretch vibration region at 3407 cm^{-1}. The observed decrease in the epoxy ring line intensity and the increase in the hydroxyl line intensity correspond to the reaction of the polymerization of epoxy resin.

The received spectral data were used for quantitative analysis of reaction kinetics in the plasma treated sample and in the control samples. The kinetics of the etching process was observed by the optical density of a stretch vibration line of aromatic ring at 1510 cm^{-1}. The form of this line does not change during etching and the aromatic ring does not take part in the reaction of polymerization. In Fig. 11.12 the dependence of the aromatic ring line intensity is presented for epoxy resin in liquid state, in cured state and during polymerization. The etching rate of cured resin is lowest, which corresponds to the highest stability of hard resin at plasma treatment. The etching rate of liquid resin is highest, which corresponds to lowest stability of liquid resin in plasma. The etching rate reflects the ability of resin macromolecules to be cut on separate low molecular parts under the action of plasma particles and to be evaporated into the vacuum of the plasma chamber. For liquid resin with initial short macromolecules the rate is highest due to the possible formation of short molecules after breaking the polymer chain under the action of plasma particles. In hard resin the macromolecules are connected by crosslinks and the appearance of the short molecules due to plasma action needs a certain amount of breaking of the polymer chain.

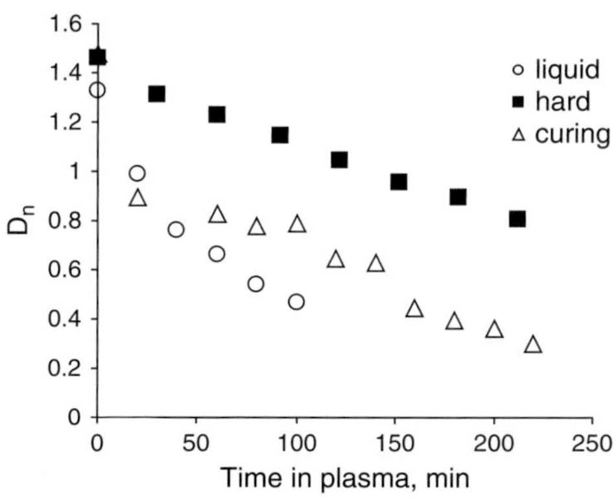

Fig. 11.12. The dependence of aromatic ring line intensity in FTIR spectra of liquid, hard and curing resins on time in plasma.

During polymerization the etching rate of the reactionable mixture decreases with time. Initially, the rate of mixture is close to the rate of liquid resin and during the polymerization reaction the etching rate of the mixture is equal to the cured resin. For the resin during polymerization the etching rate depends on the coordinate of reaction. An average rate of etching for liquid and hard epoxy resin was determined by a linear approximated curve of the optical density of the 1510 cm^{-1} aromatic ring line in FTIR spectra. On average the etching rate for liquid resin is 0.083 μm/min, for polymerizing resin the rate equals 0.040 μm/min and for hard resin the rate equals 0.038 μm/min.

The absorbance of the normalized spectra was used for analysis of chemical structure changes in epoxy resin during the reaction of polymerization under plasma action. In Fig. 11.13 the absorbance of the epoxy ring line at 915 cm^{-1} is presented for resin during curing in plasma and during curing under atmospheric conditions (control sample). In addition, the absorbance of the epoxy ring line for liquid resin without hardening under plasma action is also shown. The kinetic curve of the control sample corresponds to

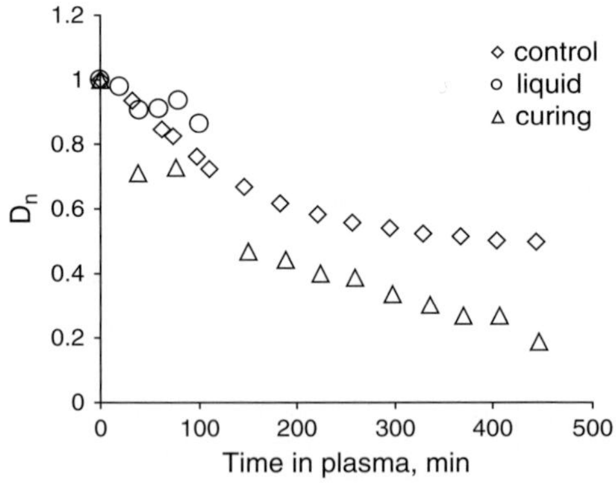

Fig. 11.13. The kinetic of curing by dependence of epoxy ring line intensity in FTIR spectra for epoxy resin with TETA hardening agent in air (rhombus), in plasma (triangles) and for liquid resin (circle) without hardening agent in plasma.

the kinetics of the second order reaction with acceleration and deceleration stages. The kinetic curve for resin cured in plasma goes deeper than the curve of the control sample. Also, some decreasing of epoxy ring line intensity is observed for liquid resin without hardening agent which is treated in plasma. The decreasing of intensity can be caused by the interaction of the epoxy ring with products of plasma destruction. The reaction of the epoxy group with free radicals formed under plasma action in the surface layer of epoxy resin is possible using the following [7]:

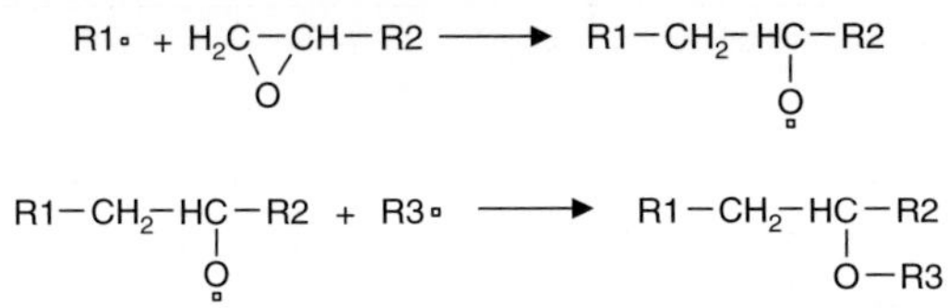

Another way is the reaction of the epoxy ring with hydroxyl and carboxyl groups of oxidized macromolecules under plasma discharge:

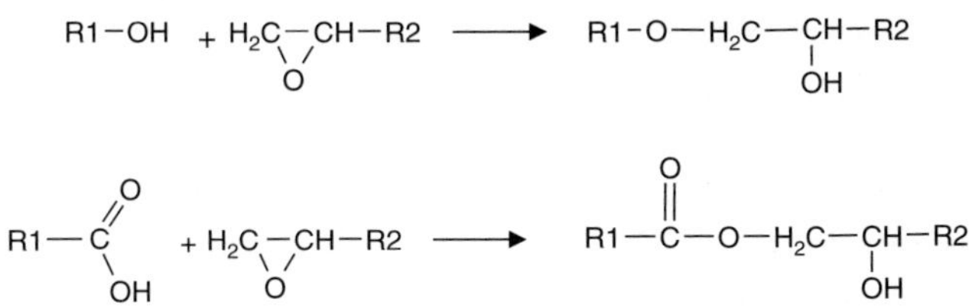

Such reactions can proceed in the epoxy resin with hardening agent under plasma treatment. Therefore, the plasma destruction products are active in curing reactions of epoxy resin which means the plasma places additional hardener into the epoxy resin. Due to additional hardener, the curing reaction proceeds deeper under plasma action than in the control sample.

The completeness of polymerization reaction in epoxy resin with TETA hardening agent was observed by Raman spectra. The Raman spectra of epoxy resin cured in air

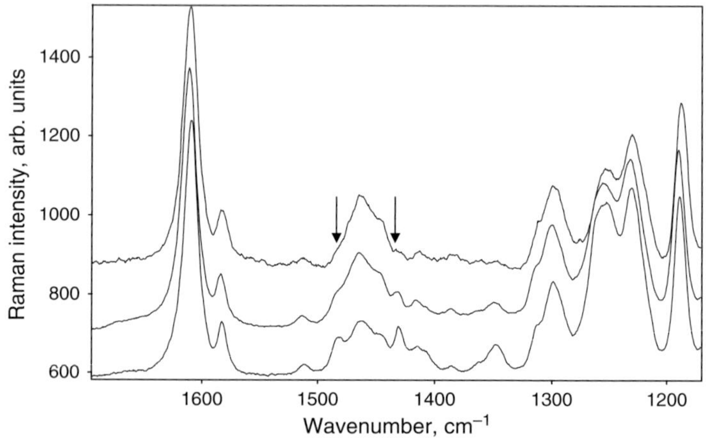

Fig. 11.14. Raman spectra of epoxy resin with TETA hardening agent: down – not cured, middle – cured in air, on top – cured in plasma. Arrays show 1482 and 1430 cm^{-1} lines.

and plasma are presented in Fig. 11.14. The lines at 1482 and 1430 cm^{-1} are observed in spectra of the initial resin. During reaction the intensity of these lines decreases which corresponds to the reaction of polymerization. In spectra of the sample cured in plasma the intensity of the 1482 and 1430 cm^{-1} lines is lower than in spectra of the sample cured in air.

In Fig. 11.15 the normalized optical density of the carbonyl group line at 1730 cm^{-1} in FTIR transmission spectra is shown for liquid, cured epoxy resin and for reactionable mixture. The absorbance of carbonyl groups in cured epoxy resin increases at initial plasma treatment and does not change during the next treatment phase. The oxidation takes place only in the thin surface layer of epoxy resin. Increasing treatment time in plasma does not lead to an increase in the carbonyl groups' concentration in epoxy resin by more than a maximal amount on the surface layer. With an increase in treatment time the carbonyl groups in the oxidized surface layer of the resin are destroyed by plasma particles during the etching process as are other parts of macromolecules in the surface layer. Therefore,

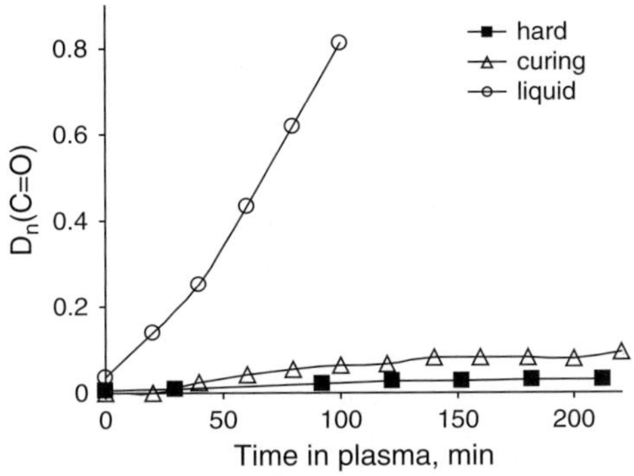

Fig. 11.15. Normalized absorbance of carbonyl group vibrations in FTIR transmittance spectra of hard, liquid epoxy resin and during curing with time in plasma.

in solid epoxy resin the formation of carbonyl groups and the etching process of the surface layer reach the dynamic equilibrium and the concentration of the carbonyl groups becomes constant.

The increase in the carbonyl groups' concentration at plasma treatment for liquid resin is significantly higher than for cured resin (Fig. 11.15). The chemical structure of cured and liquid resin does not differ and it cannot be a reason for strong oxidation instability of liquid resin in plasma. During polymerization of epoxy resin only the epoxy ring opens and the structure of the macromolecule changes insignificantly. The product of polymerization becomes weaker with the action of oxygen. The significant difference between hard and liquid resin is their viscosity.

The difference in liquid and hard resin behavior in plasma treatment can be explained by a difference in the physical state of resins. In liquid resin the processes of mixing or flow can take place during plasma treatment which is in contradiction with hard resin. Such processes can remove the oxidized layers from the treated surface of resin to the deep layer and the oxygen-containing groups cannot be reached by plasma particles for destruction. At the same time the new initial layers of resin appear on the surface and these new layers can be attacked by plasma particles as previous ones were. During plasma treatment the mixing process moves the new layers to the surface.

The intensive oxidation process of liquid epoxy resin in plasma is observed by the line at 3440 cm^{-1} of hydroxyl group vibrations (Fig. 11.16). During curing reaction under atmospheric conditions, the intensity of hydroxyl group line increases during reaction of polymerization corresponding to the appearance of hydroxyl groups as a product of the reaction. After curing the intensity of line does not change in the spectra of solid resin treated by plasma. However, in the spectra of liquid resin without hardening agent under plasma discharge the intensity of the hydroxyl group line increases in correspondence to the increase of the carbonyl line intensity. If the epoxy resin is curing in plasma discharge, the intensity of the hydroxyl group line increases more than at curing in air as it is observed for the carbonyl group line.

The treated layer due to plasma action contains a number of active centers as free radicals, oxygen-containing groups, new double bonds and others. During mixing the

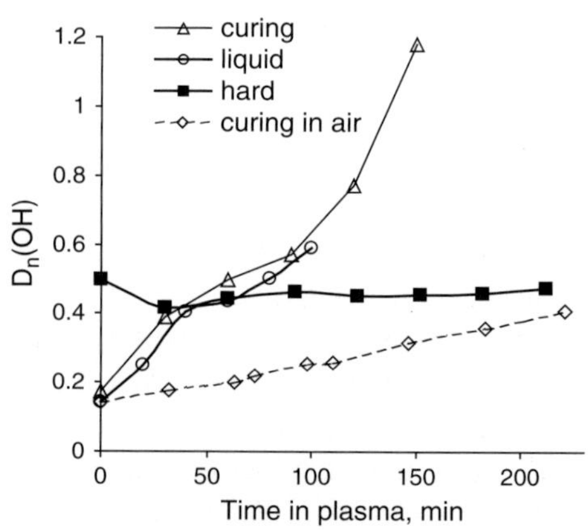

Fig. 11.16. Normalized absorbance of hydroxyl vibration line in FTIR spectra of epoxy resin with TETA hardening agent: hard resin (black cubic), liquid resin (circles), curing resin in plasma (triangles) and curing resin in air (rhombus).

treated layer is mixed with initial bulk layers of resin and the active centers can react with initial macromolecules of resin. During plasma treatment the mixing process leads to a continuous moving of treated and untreated layers. As the result of treatment the products of destruction and oxidation in bulk layers of liquid resin appear. It is observed as a strong increase in the intensity of carbonyl and hydroxyl groups' lines in transmission spectra of epoxy resin under plasma treatment which appear not only on the surface layer but also in the bulk layers.

The character of dependence of carbonyl line intensity changes with treatment time. The plasma etching of resin decreases the thickness of the sample to a low value when the mixing process cannot bring new initial layers of resin from bulk layers and the treated oxidized layers come to the surface of the resin. In this case the destruction of oxidized layers of resin becomes significant and the equilibrium of oxidation and etching processes in the surface layer is observed as for solid resin. Therefore the next increase in the oxidation of resin is impossible and the absorbance of carbonyl groups does not increase.

In the reactionable mixture of epoxy resin and hardening agent the processes of etching, oxidation and mixing take place depending on viscosity of the mixture, which becomes more viscous with curing reaction time. The increase in carbonyl group concentration in the reactionable mixture is higher than in solid resin but lower than in liquid resin. The oxidation process has extreme dependence on treatment time. The influence of mixing on the oxidation process is observed as for liquid resin. The effect of oxidation is sufficiently lower than in liquid resin due to the polymerization reaction and increase in mixture viscosity.

The morphology of the epoxy resin surface reflects the conditions of polymerization. AFM imaging does not show any defects on the surface of epoxy resin cured in vacuum (Fig. 11.17). After plasma treatment of hard epoxy resin the surface becomes rough and defects like narrow peaks appear. The same topology of the surface is observed

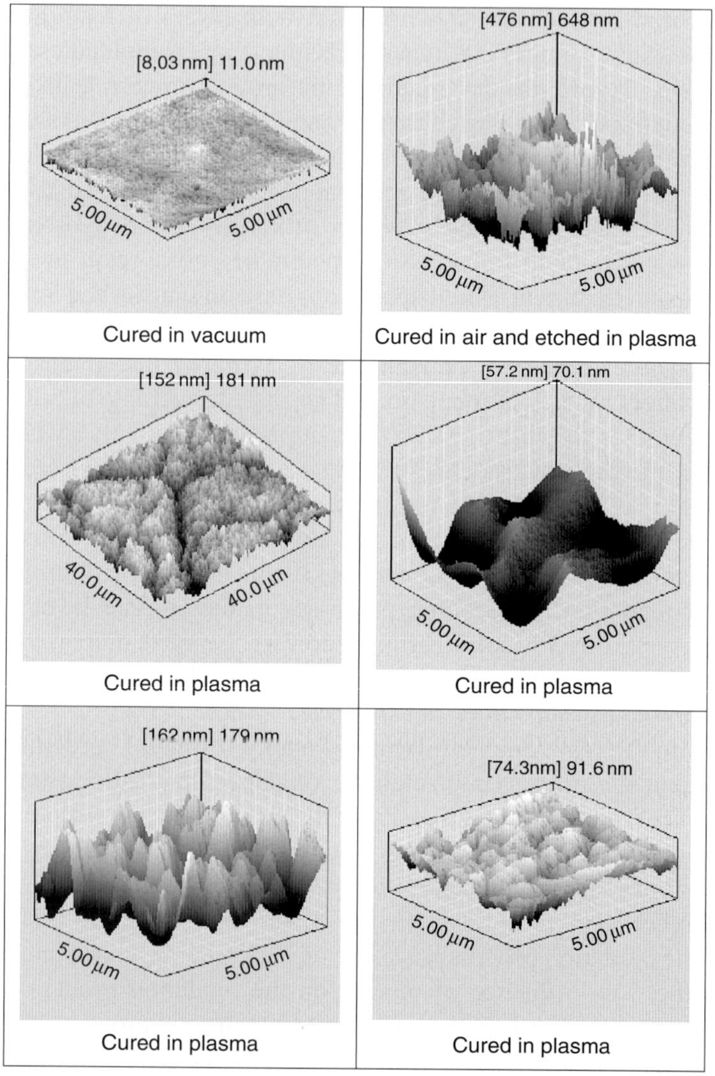

Fig. 11.17. AFM imaging of epoxy resin with TEA and TETA cured in vacuum, cured in air and then treated in plasma, and cured in plasma.

in samples of solid polymers exposed in a free space environment during space flight. Such topology corresponds to the etching process of hard material under plasma and ion beam.

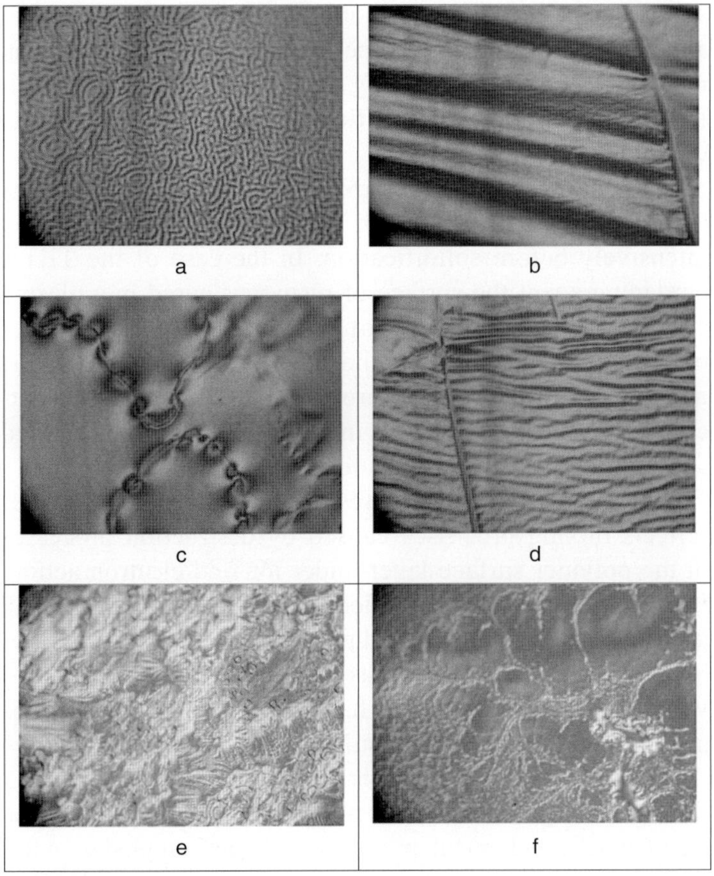

Fig. 11.18. Microphoto of epoxy resin with TEA (a–d) and TETA (e–f) cured in plasma.

The surface topology of epoxy resin cured in plasma is more complicated and contains different regions with significantly different structures. For example, there are wrinkled structures like waves (Fig. 11.18). The intensity and arrangement of wave structure are quite different. There is no clear distribution of wave structure by place, wave period or intensity. The morphology of wave structure is disturbed by lines and cracks which are distributed on the treated surface without any order or clear direction. There are surface fields with smooth surfaces like cured under atmospheric conditions. Some smooth fields contain weak narrow cracks. AFM imaging shows the detailed morphology of folded and smooth fields of surface after plasma treatment (Fig. 11.17).

Such specific surface morphology can be interpreted as the observation of the frozen flow of liquid which flowed intensively before solidification. In the case of the TETA hardening agent the surface of resin was cured in a plasma chamber and the structure of the surface was saved after treatment. In the case of the TEA hardening agent, the reactionable mixture was pre-cured in plasma and post-cured after plasma treatment in air. In all cases the wrinkled structure of the surface was observed.

The surface topology of resin cured in plasma polymer reflects mixing processes caused by destruction processes in the polymer surface layer under ion and electron action. During treatment the modified surface layer turns to bulk and the portion of unchanged polymer moves to the surface. Such a process slows down as viscosity increases and stops when the polymer becomes hard. This picture corresponds also to FTIR and Raman results shown above.

The deep structure transformations in the epoxy matrix during curing in vacuum, plasma and ion beam cause changes of mechanical properties of the composite. After curing the composite based on epoxy matrix with TEA hardening agent and glass fibers the temperature behavior of modules was analyzed by dynamical mechanical analysis (DMA). The elastic modulus G' of composite cured in air is higher than the module of composite cured after vacuum treatment and lower than the module of composite cured in plasma. The temperature behavior of the modules for samples cured in vacuum, plasma and air is different, as viewed on the G' derivative curve (Fig. 11.19). For the vacuum treated sample the phase transition (glass-transition temperature) is observed at a lower temperature than for the sample cured in air. For these samples only one phase transition is observed. The composite cured in plasma has two

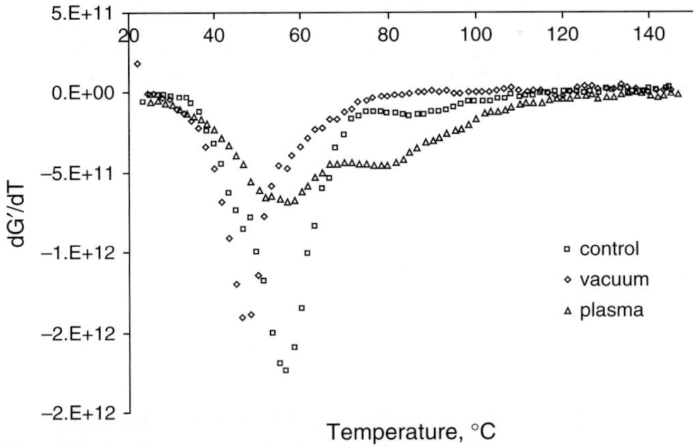

Fig. 11.19. Glass transition of composite based on epoxy resin and TEA. Cured in air, vacuum and plasma.

phase transitions. The first peak of the *G*′ derivative is at the same temperature as the sample cured in air. But the peak becomes wider and lower with intensity. The second peak is observed at a higher temperature. Compared with spectral data, the second peak corresponds to the additional polymer network formed by free radicals which were generated by plasma discharge.

References

1. V.A. Briskman, T.M. Yudina, K.G. Kostarev, A.V. Kondyurin, V.B. Leontyev, M.G. Levkovich, A.L. Mashinsky, G.S. Nechitailo, Polymerization in microgravity as a new process in space technology, Paper No. IAA-97-IAA.12.1.07, 48th International Astronautical Congress, October 6–10, 1997, Turin, Italy; *Acta Astronautica* 48(2–3) (2001), 169–180.
2. A. Kondyurin, Large size station on Mars surface by the way of polymerization of composite polymer material, Fourth Canadian Space Exploration Workshop. Science Payloads for Mars, Abstracts, Ottawa, Canada, November 15–16, 2002.

3. A. Kondyurin, Creation of biological module for self-regulating ecological system by the way of polymerization of composite materials in free space, 34th COSPAR Scientific Assembly, Houston, Texas, USA, October 10–19, Abstracts, 2002.
4. A. Kondyurin, High-size space laboratory for biological orbit experiments, 33rd Scientific Assembly of COSPAR, July 16–23, 2000, Warsaw, Poland; *Adv. Space Res.* 28(4) (2001), 665–671.
5. A.V. Kondyurin, Building the shells of large space stations by the polymerisation of epoxy composites in open space, *Int. Polym. Sci. Technol.* 25(4) (1998), T/78.
6. A. Kondyurin, G. Mesyats, Yu. Klyachkin, Creation of high-size space station by polymerization of composite materials in free space, Joint 1st Pan-Pacific Basin Workshop and 4th Japan-China Workshop on Microgravity Sciences, July 8–11, 1998, Tokyo, Japan, Paper 9-B-6; *J. Japan Soc. Microgr. Appl.* 15(Suppl. II) (1998), 61–65.
7. A. Kondyurin, B. Lauke, E. Richter, Polymerization process of epoxy matrix composites under simulated free space conditions, *High Perf. Polym.* 16 (2004), 163–175.
8. I. Kondyurina, A. Kondyurin, B. Lauke, L. Figiel, R. Vogel, U. Reuter, Polymerisation of composite materials in space environment for development of a Moon base, *Adv. Space Res.* 37 (2006), 109–115.
9. M.G. Kroshkin, *Physical-chemical Bases of Space Studies*, Mashinostroenie, Moscow, 1969.
10. A. Kondyurin, K. Kostarev, M. Bagara, Polymerization processes of epoxy plastic in simulated free space conditions, Paper IAF-99-I.5.04, 50th International Astronautical Congress, October 4–8, 1999, Amsterdam, The Netherlands; *Acta Astronautica* 48(2–3) (2001), 109–113.
11. H.U. Walter, *Fluid Sciences and Materials Science in Space. A European Perspective*, Springer-Verlag, Berlin, 1987.
12. T.F. Klein, G.A. Lesieutre, Space environment effects on damping of polymer matrix carbon fiber composites, *J. Spacecr. Rock.* 37(4) (2000), 519–525.
13. K.K. de Groh, B.A. Banks, A.M. Hammerstrom, E.E. Youngstrom, C. Kaminski, L.M. Marx, E.S. Fine, J.D. Gummow, D. Wright, MISSE PEACE polymers: an International Space Station environmental exposure experiment,

Proceedings of the Conference on ISS Utilization – 200, Cape Canaveral, Fl, AIAA 2001-4923; also appears in NASA TM-2001-211311, October 15–18, 2001.

14. H.G. Pippin, *Final Report on Analysis of Boeing Specimens Flown on the Effects of Space Environment on Materials Experiment*, Boeing Phantom Works, 1999.
15. J.W. Connell, The effects of low-Earth orbit atomic oxygen exposure on phenylphosphine oxide-containing polymers, Final report, Evaluation of Space Environment and Effects on Materials (ESEM), Appendix D.
16. R.L. Kiefer, R.A. Orwold, J.E. Harrison, V.M. Ronesi, S.A. Thibeault, The effects of the space environment on polyetherimide films, Final report, Evaluation of Space Environment and Effects on Materials (ESEM), Appendix C.
17. B. Czaubon, A. Paillos, J. Siffre, R. Thomas, Mass spectrometric analysis of reaction products of fast oxygen atoms – material interactions, *J. Spacecr. Rock.* 35(6) (1998), 797–804.
18. ECSS Space Environment Standard, ECSS E-10-04 (Guide for LEO mission), ECSS-Q-70-04 (outgassing), ESA, 2000.
19. S. Koontz, K. Albyn, L. Leger, Atomic oxygen testing with thermal atom systems: a critical evaluation, *J. Spacecr.* 28(3) (1991), 315–323.
20. S. Koontz, L. Leger, K. Albyn, J. Cross, Vacuum ultraviolet radiation/atomic oxygen synergism in materials reactivity, *J. Spacecr.* 27(3) (1989), 346–348.
21. F. Lura, D. Hagelschuler, V.V. Abraimov, *The Complex Simulation of Essential Space Environment Factors for the Investigation of Materials and Surfaces for Space Applications*, DLR, Berlin, Germany.
22. K.K. de Groh, M. Morgana, The effect of heating on the degradation of ground laboratory and space irradiated Teflon FEP, NASA TM-2002-211704.
23. S.T. Lai, D.J. Della-Rose, Spacecraft charging at geosynchronous altitudes: new evidence of existence of critical temperature, *J. Spacecr. Rock.* 38(6) (2001), 922–928.
24. C.K. Purvis, H.B. Garrett, A.C. Whittlesey, N.J. Stevens, Design guidelines for assessing and controlling spacecraft charging effects, NASA TP-2361, 1984.
25. D.P. Cadogan, S.E. Scarborough, Rigidizable materials for use in gossamer space inflatable structures, AIAA-2001-1417.

26. M. Iwata, A. Ohnishi, H. Hirosawa, F. Tohyama, Measurement and evaluation of thermal control material with polyimide for space use, *J. Spacecr. Rock.* 38(4) (2001), 504–509.
27. O.N. Favorskii, Ya.S. Kadaner, *About Heat Transfer in Space*, Visshaya Shkola, Moscow, 1972.
28. L.A. Teichman, W.S. Slemp, W.G. Witte, Jr, Evaluation of selected thermal control coatings for long-life space structures, NASA TM-4319, 1992.
29. E.A. Barbashev, M.I. Dushin, Yu.N. Ivonin, V.I. Kozin, E.F. Nikishin, B.I. Panshin, B.V. Perov, Some results of tests of polymer materials after exposition in conditions of free space, in *Space Technology and Material Science*, Moscow, Nauka, 1982.
30. J.H. Fu, G.R. Graves, Thermal environments for space shuttle payloads, in *AIAA Shuttle Environment and Operation II Conference Proceedings*, 1985, p. 18.
31. Y. Haruvy, Risk assessment of atomic-oxygen-effected surface erosion and induced outgassing of polymeric materials in LOE space systems, *ESA Journal*, 14 (1990), 109–119.
32. J. Dever, C. Semmel, D. Edwards, R. Messer, W. Peters, A. Carter, D. Puckett, Radiation durability of candidate polymer films for the next generation space telescope sunshield, NASA TM 2002-211508 and AIAA-2002-1564.
33. V.P. Begishev, O.Y. Smetannikov, N.A. Trufanov, I.N. Shardakov, *Int. J. Polym. Sci. Technol.* 25 (1998), T85.
34. J.W. Vanderhoff, M.S. El-Aasser, F.J. Micale *et al.*, Preparation of large particle-size monodisperse latexes in space, *PMSE Proc. Am. Chem. Soc., Div. Polym. Mat. Sci. Eng.* 54 (1986), 587.
35. A. Sadykov, V. Leontyev, Y. Mangutova, G. Grechko, G. Nechitailo, A. Mashinsky, Systems of chemical reactions depending on gravity, *Akademiia Nauk SSSR, Doklady 303* (1988), 1004.
36. G.S. Nechitailo, A.L. Mashinsky, *Space Biology. Studies at Orbital Stations, Mir*, Moscow, 1993, 503 pp.
37. L. Bogatyreva, V. Briskman, M. Levkovich, T. Lyubimova *et al.*, *Space Sci. Technol.* 1989, 43.
38. R. Muller, D. Sturm, H.J. Rath, Photoinitiated radical polymerisation of liquid monomers in microgravity, *Proc. VIII Europ. Symp. on Material and Fluid Sciences in Microgravity*, ESA SP-333, V.2, 1992, 895.

39. J.A. Pojman, D.I. Fortenberry, V.M. Ilyashenko, Frontal polymerisation as an analog of SHS, *Int. J. Self-Prop. High Temp. Synth.* 6(3) (1997), 355.
40. V.A. Briskman, K.G. Kostarev, T.P. Lyubimova, V.L. Levtov, V.V. Romanov, Polymerization under microgravity: results and prospects, *Cos. Res.* 39(4) (2001), 338–350.
41. V.A. Briskman, K.G. Kostarev, A.V. Shmyrov, High-gravity as a research tool in studying the nature of structure formation in polymers, *Adv. Space Res.* 29(4) (2002), 599–607.
42. K.G. Kostarev, T.M. Yudina, Frontal polymerization under convective motion of reaction mixture, *RSC*, in press.
43. K.G. Kostarev, T.M. Yudina, and N.V. Pistsov The effect of the free convection on the poly(acrylamide) gel structure formation and properties, *Polym. Sci., Ser. A*, 42(11) (2000), 1248–1254.
44. C.-H. Lee, L.W. Chen, Reactive probability of atomic oxygen with material surfaces in low Earth orbit, *J. Spacecr. Rock.* 37(2) (2000), 252–256.
45. E. Grossman, Y. Lifshitz, J.T. Wolan, C.K. Mount, G.B. Hoflund, In situ erosion study of Kapton using novel hyperthermal oxygen atom source, *J. Spacecr. Rock.* 36(1) (1999), 75–78.
46. R.I. Gonzales, S.H. Phillips, G.B. Hoflund, In situ oxygen atom erosion study of polyhedral oligomeric silsesquioxane-siloxane copolymer, *J. Spacecr. Rock.* 37(4) (2000), 463–467.
47. M.A. Golub, T. Wydeven, Reactions of atomic oxygen (O(3P)) with various polymer films, *Polym. Degrad. Stab.* 22 (1988), 325–338.
48. B.A. Banks, R. Demko, Atomic oxygen protection of materials in low Earth orbit, Presented at the 2002 Symposium, NASA TM 2002-211360, February 2002.
49. B.A. Banks, K.K. de Groh, B.M. Auer, L. Gebauer, C. LaMouraux, Atomic oxygen interaction at defect sites in protective coatings on polymers flown on LDEF, LDEF Materials Results for Spacecraft Applications, NASA CP-3257, 1994, pp. 143–158.
50. B.A. Banks, K.K. de Groh, J.L. Bucholz, M.R. Cales, Atomic oxygen interactions with protected organic materials on the Long Duration Exposure Facility (LDEF), LDEF-69 Months

in Space, Third Post-Retrieval Symposium, NASA CP-3275, Part 2, 1993, pp. 737–753.
51. B. Banks, K.K. de Groh, S.K. Rutledge, C.A. Haytas, Consequences of atomic oxygen interaction with silicone and silicone contamination on surfaces in low Earth orbit, NASA report TM-1999-209179.
52. B.A. Banks, A. Synder, S.K. Miller, R. Demko, Issues and consequences of atomic oxygen undercutting of protected polymers in low Earth orbit, Presented at the Sixth International Conference on Protection of Materials and Structures from Space Environment, Toronto, Canada, NASA TM-2002-211577, May 1–3, 2002.
53. J.B. Cross, S.L. Koontz, J.C. Gregory, M.J. Edgell, Hyperthermal atomic oxygen reactions with Kapton and polyethylene, in *Materials Degradation in Low Earth Orbit (LEO)*, Edited by V. Srinivasan and B.A. Banks, The Minerals, Metals and Materials Society, 1990.
54. J.R. Cruz, C.H. Shah, A.S. Postyn, Properties of two carbon composite materials using LTM25 epoxy resin, NASA TM-110286, 1996.
55. J.A. Dever, K.K. de Groh, Vacuum ultraviolet radiation and atomic oxygen durability evaluation of HST Bi-Stem Boom thermal shield materials, NASA TM-2002-211364, February 2002.
56. J.R. Gaier, M.L. Davidson, R. Shively, Durability of intercalated graphite epoxy composites in low Earth orbit, NASA TM-107157, March 1996.
57. A. Gindulyte, L. Massa, B.A. Banks, S. Miller, Direct C—C bond breaking in the reaction of O(P) with fluoropolymers in low Earth orbit, *J. Phys. Chem.* article; *ASAP* article, May 3, 2002.
58. M.A. Golub, T. Wydeven, R.D. Cormia, ESCA study of several fluorocarbon polymers exposed to atomic oxygen in low Earth orbit within or downstream from a radio-frequency oxygen plasma, *Polymer* 30 (1989), 1571–1575.
59. K.K. de Groh, B.A. Banks, Atomic oxygen undercutting of long duration exposure facility aluminized Kapton multilayer insulation, *J. Spacecr. Rock.* 31(4) (1994), 656–664.
60. K.K. de Groh, B.A. Banks, G.W. Clark, A.M. Hammerstrom, E.E. Youngstrom, C. Kaminski, E.S. Fine, L.M. Marx, A sensitive technique using atomic force microscopy to measure the

low Earth orbit atomic oxygen erosion of polymers, Presented at the Poly Millennial 2000 Conference, Kona, HI, December 9–13, 2000; also appears in NASA TM-2001-211346 (December 2001).
61. K.K. de Groh, B.A. Banks, R. Demko, Techniques for measuring low Earth orbital atomic oxygen erosion of polymers, *Proceedings of the SAMPE 2002 Conference*, Long Beach, CA, pp. 1279–1292, May 6–10, 2002; NASA TM-2002-211479.
62. M.L. Illingsworth, B.A. Banks, J.W. Smith, D. Jayne, R.G. Garlick, S.K. Rutledge, K.K. de Groh, Plasma and beam facility atomic oxygen erosion of a transition metal complex, *Plasma Chem. Plasma Proc.* 16(1) (1996).
63. Z.A. Iskanderova, J.I. Kleiman, Yu. Gudimenko, R.C. Tennyson, Influence of content and structure of hydrocarbon polymers on erosion by atomic oxygen, *J. Spacecr. Rock.* 32(5) (1995), 878–884.
64. D.A. Jaworske, K.K. de Groh, G.M. Podojil, T.A. McCollum, J. Anzic, Leveling coatings for reducing atomic oxygen defect density in graphite fiber-epoxy composites, *J. IES* (May/June 1994), 26–31.
65. M.C. Lillis, E.E. Youngstrom, L.M. Marx, A.M. Hammerstrom, K.D. Finefrock, C.A. Youngstrom, C. Kaminski, E.S. Fine, P.K. Hunt, K.K. de Groh, B.A. Banks, E.A. Sechkar, Space flight experiments to measure polymer erosion and contamination on spacecraft, NASA TM-2002-211553, June 2002.
66. J.-F. Roussel, A. Bourdon, Oxygen interaction with materials III: data interpretation via computer simulation, *J. Spacecr. Rock.* 37(3) (2000), 324–330.
67. S.K. Rutledge, B.A. Banks, A technique for synergistic atomic oxygen and vacuum ultraviolet radiation durability evaluation of materials for use in low Earth orbit, Prepared for the 3rd International Conference on Protection of Materials and Structures from the Low Earth Orbit Space Environment cosponsored by the Canadian Space Agency and the Institute for Space and Terrestrial Studies, Toronto, Canada, April 25–26, 1996, NASA-TM-107230.
68. S.K. Rutledge, B.A. Banks, M. Kitral, A comparison of space and ground based facility environmental effects for FEP Teflon, NASA report TM-1998-207918/rev1.

69. S.K. Rutledge, R.M. Olle, Space Station Freedom solar array blanket coverlay atomic oxygen durability testing results, Prepared for the 38th International SAMPE Symposium, Anahiem, California, May 10–13, 1993.
70. A. Snyder, Investigation of atomic oxygen erosion of polyimide Kapton H exposed to a plasma asher environment, NASA report TM-1999-209178.
71. A. Snyder, B.A. Banks, Fast three-dimensional method of modelling atomic oxygen undercutting of protected polymers, Presented at the Sixth International Conference on Protection of Materials and Structures from the Space Environment, NASA 2002-211578, Toronto, Canada, May 1–3, 2002.
72. T.J. Stueber, C. Mundson, Evaluation of Kapton pyrolysis, arc tracking, and flashover on SiOx-coated polyimide insulated samples of flat flexible current carriers for SSF, NASA CR-191106, April 1993.
73. K. Yokota, M. Tagawa, N. Ohmae, Impingement angle dependence of erosion rate of polyimide in atomic oxygen exposures, J. Spacecr. Rock. 39(1) (2001), 155–156.
74. T.M. Zuby, K.K. de Groh, D.C. Smith, Degradation of FEP thermal control materials returned from the Hubble Space Telescope, NASA TM-104627, 1995.

Index